Praise for the fir~

This is a testament to the brave ⸺ ‚viwa. His words are an inspiration to anyone fig₁ ⸺ ‚y, and a reminder to oppressors the world over that the ₁. ⸺ ‚an never be broken. – **Noo Saro-Wiwa**, author of *Looking for Tran‗ ‗erland: Travels in Nigeria* (2012).

Here is a remarkable book of the correspondence from one of the greatest leaders of our time to a strong and gentle Catholic sister living half-way around the world. Ken Saro-Wiwa, enduring harsh treatment and facing certain death, writes from detention in Nigeria about justice and honour and sets the bar for courage for the rest of us. Struggles for indigenous justice in the face of corporate tyranny continue to this day. Everyone engaged in these struggles will be moved and inspired by these haunting letters written by a legend. – **Maude Barlow**, author, activist and National Chairperson of Council of Canadians.

A poingnant collection that unveils a remarkable friendship as much as it animates the memory of Saro-Wiwa's indomitable spirit. It is perhaps one of the the bitter ironies of his life he had to feed the soldiers who gurarded him as well as witness army captains fight over who should be his jailer. – **Brian Chikwava**, write and winner of the 2004 *Caine Prize for African Writing*, Associate Editor, *Wasafiri Magazine*

More fully than any biographical essay would have done, the letters and the poems reveal the mind of the campaigner for justice while he is under arrest, courageously planning and prompting, writing and keeping himself informed, keeping his cause alive, but they also show Ken Saro-Wiwa as the anxious father worrying about his children and as the man alone thrown on his resources. The three lucid essays which frame the letters prove an excellent and informative guide to the events behind the letters and add to the imporance of this publication. – **Abdulrazak Gurnah**, novelist, Booker Prize nominee 2004, winner of the RFI *Témoin du Monde Prize* 2006 and Professor of English at the University of Kent

The letters and poems collected in this volume show with great eloquence that Saro-Wiwa confronted Abacha's darkness, and the darkness of the international oil conglomerates, especially Shell, with anger, sadness, wit and humour. In nearly every letter and poem in the volume there is suffusing light and uncommon grace. I confidently expect that in time, this slim volume will take its rightful place among the most important works of prison writing and environmental activism in the world. – **Biodun Jeyifo**, Harvard University

Following Ken Saro-Wiwa's second arrest in 1994, Sr Majella McCarron approached Trócaire for help. His release became a priority campaign for us, and we engaged with Shell, the media and Government to try and commute the death sentences for him and the eight co-accused Ogoni leaders. I remember the despair in Trócaire's offices on 10th November 1995 when we learned that all nine had been executed. The struggle of the Ogoni people is a part of Trócaire's history, and the writings in *Silence Would be Treason: Last Writings of Ken Saro-Wiwa* are a testament to Ken Saro-Wiwa's spirit and courage, demonstrating that, even in the darkest of times, love truly can conquer fear. – **Éamonn Meehan**, Executive Director of Trócaire

I couldn't help but wonder what took this book so long to come out. Perhaps because of the medium–personal letters to Sr. Majella–yet still, given their scope: the political content and indomitable spirit, the environmental issues in Ogoniland, the international campaign for the Ogoni Nine and the prospects for democracy in Nigeria and elsewhere, one would have hoped that it was clear they belonged in the public domain. Considering the delay therefore, 18 years of silence have been committed. Thank goodness the book is now available and what a story! The kind that would not forgive silence if it had remained untold. – **Mildred Barya**, *African Literary News*, 2013

Clear and direct, these letters and poems are the last expression of a voice the regime was determined to silence: a voice for indigenous rights, environmental survival and democracy, many of those battles were won despite his death and whose voice comes alive today again in these extraordinary letters. – *Boletim Africanista*, 2013

Silence Would be Treason – *Last Writings of Ken Saro-Wiwa*, is a great book. It revives and supplements the fading memories of actors and actresses like us (not spectators) during the gloomy days. It needs to be read carefully with an open mind. The book contains correct information about the hey days of the Ogoni struggle, its victories, failures, betrayals and travails in the naked face of highly organized state/corporate violence and conspiracies against a marginalised and embittered people of the eastern Niger Delta belt in Nigeria. – **Patrick Naagbanton**, Port Harcourt, Nigeria, 2014

Silence Would Be Treason

Silence Would Be Treason

Last writings of Ken Saro-Wiwa

*ÍDE CORLEY, HELEN FALLON, AND
LAURENCE COX*

DARAJA PRESS

Originally co-published by CODESRIA (http://www.codesria.org) and Daraja Press
(https://darajapress.com) 2014. This new edition is published by Daraja Press.
Text letters and images: Copyright © 2018 John Paul II Library, Maynooth University
Ken Saro-Wiwa's poems: ©2018 Ken Saro-Wiwa Estate.

Library and Archives Canada Cataloguing in Publication

Saro-Wiwa, Ken, 1941-1995
[Works. Selections]
Silence would be treason : last writings of Ken Saro-Wiwa /
Íde Corley, Helen Fallon, Laurence Cox, editors. – Second edition.

Originally published: Dakar, Senegal : The Council for the Development of
Social Science Research in Africa ; Nairobi : Daraja Press, 2013.
Includes bibliographical references and index.
ISBN 978-1-988832-24-1 (softcover)

1. Saro-Wiwa, Ken, 1941-1995. 2. Saro-Wiwa, Ken, 1941-1995–
Correspondence. 3. Nigerian poetry (English). I. Corley, Íde, editor
II. Fallon, Helen, editor III. Cox, Laurence, editor IV. Title.

PR9387.9.S27A6   2018 828'.91409   C2018-904323-7

But while the land is ravaged
And our pure air poisoned
When streams choke with pollution
Silence would be treason...

Ken Saro-Wiwa

Contents

Ken Saro-Wiwa's Letters to Sister Majella

Poems

Preface to new edition

NOO SARO-WIWA

IN 1992, WHEN I WAS A TEENAGER, my father wrote me a letter from Nigeria informing me that his campaign for environmental and human rights was intensifying and that the government could kill him. He hadn't been incarcerated at that point; life seemed normal and I read his words with incredulity. Surely he was scaremongering, the way parents sometimes do. How could anyone contemplate their own violent demise so collectedly?

It is said that courage is defined not by the absence of fear but the ability to overcome it. My father knew the risks when he took on the Nigerian military government and Shell Oil. The fact that he could go to battle, eyes wide open, against such formidable opponents was a mark of his hardiness and ambition. We the Ogoni people were specks against the giant rockface of Nigeria's military-industrial complex, and until the 1990s few outside the Niger Delta region knew who we were. Yet my father brought our environmental problems to the planet's consciousness through persistence and a belief that justice will eventually prevail when pursued peacefully.

The letters he wrote to Sister Majella from his prison cell reveal a lack of self-pity. He focused on family, community and the struggle, making no distinction between his own fortunes and that of the Ogoni and Nigeria. Success at an individual level was meaningless to him unless shared by everyone.

His sacrifice is a lesson and inspiration to us all.

Noo Saro-Wiwa
London, 2018

About the contributors

Nnimmo Bassey coordinates Oilwatch International and also directs the Health of Mother Earth Foundation—an environmental justice think tank. He served as the executive director, Environmental Rights Action/ Friends of the Earth, Nigeria, 1993-2013 and chaired Friends of the Earth International, 2008-2012. Bassey has authored books on the environment, architecture, management and poetry. His poetry collections include *We Thought It Was Oil But It Was Blood* (2002) and *I Will Not Dance to Your Beat* (2011). *To Cook a Continent: Destructive Extraction and the Climate Crisis in Africa* (2012) published by Pambazuka Press. His latest book, *Oil Politics: Echoes of Ecological Wars* is published by Daraja Press. He was named as one of Time magazine's Heroes of the Environment in 2009 and was a co-recipient of the 2010 Right Livelihood Award also known as the "Alternative Nobel Prize." In 2012 he received the Rafto Human Rights Award.

Íde Corley is a Lecturer in English at Maynooth University where she directs MA programmes in Postcolonial and World Literatures and in Irish Literature and Culture. Her research focuses primarily on twentieth-century literary engagements with the politics of black unity associated with tri-continental Pan-African nationalism, African socialism and modern African identity. She was the Principal Convenor of an ESF-funded workshop entitled "Multiple Modernities of Same-Sex Sexuality in Nigeria" in 2010 and has published articles and reviews in *Modern Language Studies, Interventions* and *Journal of Postcolonial Writing.*

Laurence Cox co-directs the MA in Community Education, Equality and Social Activism at Maynooth University. He is cofounder of the international, open-access social movement journal *Interface*, co-editor of *Understanding European Movements: New Social Movements, Global Justice Struggles, Anti-Austerity Protest, Marxism and social Movements* and author of *Buddhism and Ireland*. He is currently part of an international team researching the life of U Dhammaloka, an Irish migrant worker who became a Buddhist monk and anti-colonial activist in early 20th Century Burma. Dr Cox has been involved in a wide range of social movements in several countries for over quarter of a century.

Mark Dummett is Business and Human Rights researcher at Amnesty International, based in London. He has investigated human rights violations linked to business activities in Nigeria, the Democratic Republic of Congo, Myanmar and Bangladesh. On Nigeria, Mark and his colleagues have focussed on two areas: the failure of oil companies, including the UK-Dutch multinational Shell, to prevent and adequately clean up oil spills that destroy the environment of the Niger Delta; and secondly, the role that Shell played in the 1990's during the military crackdown on protests in Ogoni. Before joining Amnesty, Mark worked as a correspondent for the BBC.

Helen Fallon is Deputy University Librarian at Maynooth University. She has worked in libraries in Sierra Leone, Tanzania, Namibia and Saudi Arabia. Her professional interests include libraries in developing countries, African women writers, staff development, academic publishing, creativity, and the leadership and marketing of academic libraries. She has published extensively and runs workshops on academic publishing and maintains a blog for library staff who wish to write for publication at http://academicwritinglibrarian.blogspot.ie/

Sister Majella McCarron was born in Derrylin, County Fermanagh. She joined the Missionary Institute of Our Lady of Apostles in 1956. After graduating with a science degree from University College Cork, she taught for thirty years in Nigeria, in secondary schools and at the University of Lagos. She worked closely with Ken Saro-Wiwa on issues of justice and the environment and was compelled to campaign for the lives of the Ogoni 9, hanged on the 10th of November 1995. She now lives in Ireland, continuing her work on environmental justice. In 2011, she donated letters she received from Ken Saro-Wiwa to the Library at Maynooth University.

Graham Kay is a PhD candidate and John and Pat Hume Scholar with the Department of History at Maynooth University. In 2015, he was the inaugural recipient of the Ken Saro-Wiwa bursary. His thesis, entitled: 'The First Oil War: Great Britain, Germany, and the race for oil, 1896-1921' is the main focus of his research. Graham completed his BA in History at Maynooth and his MA in War Studies at King's College, London. His research interests include: contemporary security, cyber-security, and transnational comparative history. Graham lectures occasionally on these topics through the Centre of Military History and Strategic Studies at Maynooth University. You can contact Graham by email at: grahammkay@gmail.com and via Twitter: @GrahamMKay

Dr. Anne O' Brien is a lecturer with the Department of Media Studies at Maynooth University. She has published a number of articles on the representation of women in radio and television, on women workers in screen production industries and examined on-going gender inequality in media production and representation. She has also undertaken research on community media, examining its social benefit and governance needs. Her book, *The Politics of Tourism Development, Booms and Busts in Ireland* (Palgrave Macmillan, 2011) examines the role of the state in the development of Irish tourism. She has produced a number of documentaries and series for Irish national broadcasters, RTE and TG4 as well as the Ken Saro-Wiwa audio archive for Maynooth University.

Noo Saro-Wiwa is the daughter of Ken Saro-Wiwa. Born in Port Harcourt, Nigeria, and raised in England, she is an author and journalist. Her first book, *Looking for Transwonderland: Travels in Nigeria* (Granta, 2012), was selected as BBC Radio 4's Book of the Week in 2012 and named *The Sunday Times Travel Book of the Year*, 2012. It has been translated into French and Italian.

Acknowledgments

OUR THANKS FIRST OF ALL to Sr. Majella McCarron for this generous donation which reflects so much of her own life's work. Thanks too to John O'Shea
who formed the connection between Sr. Majella and Maynooth University. In
Maynooth University, thanks are due to Cathal McCauley, University Librarian, for much support with the project and to all the library staff who assisted
in various ways. Sincere thanks to Dr Ciara Gallagher, postdoctoral scholar
in the English Department, for transcription and other work on the original
manuscript. The publication of the first edition of this book was made possible with the support of Maynooth University and Trócaire.

Thanks also to the Saro-Wiwa family, in particular Owens Wiwa, the late Ken
Wiwa and Noo Saro-Wiwa for their support for this project.

Foreword to the first edition

NNIMMO BASSEY

THESE LETTERS AND POEMS are invaluable fragments of a living conversation that portrays the indomitable power in humans to stay alive in the face of certain death – to stay alive even in death. Reading through the treasure trove of the letters and poems compiled here as *The Last Writings of Ken Saro-Wiwa* evoked such intense memories of his resolute struggles against an oil behemoth and a deaf autocratic government. His crusade frames one of the most tumultuous periods of Nigeria's history; his tragic story evokes anger and demands action to resolve the crises that first led the Ogoni people to demand that Shell clean up Ogoni or clear out of the territory. It was his leadership, in great part, that forced Shell out of Ogoni in January 1993.

These letters are a testament of hope. Being one side of robust conversations between two persons that many would find unlikely as close friends, we learn the lessons that indeed 'friends love at all times and brothers (and sisters) are born for adversity', as a proverb in the Bible states. This is where we must applaud Sister Majella McCarron for preserving and making public these letters that Ken Saro-Wiwa wrote to her between 20 October 1993 and 14 September 1995.

As I read through this collection I felt a sense of awe at the positive disposition of this saint–as aptly described by his son, Ken Wiwa Junior. The harshness of his prison conditions could not kill his strong sense of duty to his people and indeed, as the reader will find, his acerbic sense of humour. Never did the pressures of the tyrannical government dent his vision for an Ogoni free of pollution, oppression and exploitation. He worked assiduously to secure the dignity of the Ogoni and by extension, I would say, of other oppressed Nigerians and of all humanity.

The letters reveal to us a well-rounded man. His love for his family and concern for each individual in that close circle come through very strongly. His love for the Ogoni people and his readiness to invest as many resources in the struggle as he could muster must challenge our commitment to the causes that we support. His vulnerable position comes through when he states, in one of the letters, "I'm not particularly protected, although I have great faith in God, in the justness of my cause & in the belief in eventual VICTORY. But the pain which we all have to endure! Would to God it had been lighter!"[1] In the same letter he states, "But I'm in good spirit, undaunted, as convinced of my cause as ever. My real worry is the devastation of Ogoni villages, the destabilization of the area & the harassment & killing of the people." This letter tells us much more: "I'm not worried for myself. When I undertook to confront Shell & the Nigerian establishment, I signed my death warrant, so to speak. At 52, I think I've served my time and, come to face it, I've lived a charmed life. A few more books, maybe, & the opportunity to assist others would have been welcome. But it's okay."

The letters reveal to us that being a keen student of current history helped to keep him in high spirits while held in the dungeon. His letter of 29/10/94 informs, "...do not forget that I've been here only 23 weeks now. Mandela & Sisulu were there for 26/27 years. How can I complain?"[2]

Repeatedly, Saro-Wiwa bemoaned the fact that MOSOP (Movement for Survival of the Ogoni People) had not adequately prepared the generality of the Ogoni people for the struggle. This is a signal for all who are engaged in mass mobilisations and movement building.

At a time when my nation is mired in extreme cases of corruption and where security is assured mainly for those who follow the deviant path, we see in Ken Saro-Wiwa a different breed—a man who did not hesitate to spend his resources but was also keen to stay transparent and accountable to his people. Could it be that this irked the corporate and state gangs that relentlessly pursued him and eventually dropped the noose around his neck?

A man of letters, no pun intended, Ken Saro-Wiwa challenges us in his strong eye for details. His care to check grammatical errors in documents prepared by others, even while in the *gulag*, is stunning. Why would a man, literally in the belly of the beast, care about spelling errors in other people's writings, in statements—even newspaper reports?

When Wale Okediran reviewed Ken Saro-Wiwa's book in *On a Darkling Plain* in 1993 and suggested that he was too hard on the erstwhile leader of Biafra, Saro-Wiwa responded in a letter to Okediran that in "considering what Ojukwu had done to my people the Ogonis, I was actually very mild with him." According to Okediran, what really surprised him when he met Saro-Wiwa in person was that "he had the effrontery to point out some grammatical mistakes he said I made in the review." He acknowledges that

1. MU archive PP/7/4
2. MU archive PP/7/14

Saro-Wiwa was "never shy of controversy, ever ready for a good argument and always finical with tiresome editorial details like punctuation marks and typos. Kenule Beeson Saro-Wiwa is one hell of a human maelstrom who leaves you breathless with his stamina and dexterity for discussions on every conceivable subject under the sky."[3]

The fact that he was concerned with details and aimed for perfection in his own writing can be seen in his requests to McCarron to help edit some of his drafts, such as his acceptance speech when he won the Right Livelihood Award—also known as the Alternative Nobel Prize— in 1994.

Ken Saro-Wiwa and Sister McCarron both influenced my life and growth as an environmental justice advocate. In addition, Saro-Wiwa challenged me as a fledging writer who thought I would find a niche as a poet and short story writer. His pioneering work in building a virile environmental justice movement as well as those for the rights of minorities in Nigeria remains outstanding and continues to inspire campaigners around the world. I recall his visit to my humble home in Benin City when he came to lead a conference of the Association of Nigerian Authors (ANA) in 1994. It was a memorable occasion and I heartily drank from his spring of wisdom on a variety of topics. We veered to issues of the rights of minorities in Nigeria. When I made the claim that I was also from a minority ethnic nation, he laughed and stated firmly that I could not be right because, with a population of close to three million, the Ibibio could not claim the minorities tag. Arguments on that led us nowhere and I knew better than to get trapped in it. There was so much to learn, and, as it turned out, there was very little time left.

November 10, 1995: that date has come to stay with us. It is a date that is both frozen and unforgettable. On that day, few gave any thought to the Port Harcourt Prisons and the diabolical plans of General Abacha and his henchmen. The Association of Nigerian Authors (ANA) was holding its annual conference at the University of Lagos, and by this time I was the General Secretary and Odia Ofeimum was the President. Saro-Wiwa's tenure as ANA president ended in 1993. One of the issues of critical concern at that 1995 conference was the statement ANA was to issue concerning the death sentence passed by the kangaroo tribunal that had tried Ken Saro-Wiwa, Saturday Dobee, Nordu Eawo, Daniel Gbooko, Paul Levera, Felix Nuate, Baribor Bera, Barinem Kiobel, and John Kpuine.

---

3. Wale Okediran, "Ken Saro-Wiwa at 50". *Guardian* (London). 19th October 1991, p. 11

We debated over the tenor of the statement. Should ANA denounce the arrest, detention and sham trial/verdict passed on its past president, Ken Saro-Wiwa, or should the statement be a politely worded plea for leniency, with nothing said to ruffle the feathers of the dark-shaded despot, Abacha? It was while we were locked in this debate that the news filtered through that Ken Saro-Wiwa and the other eight Ogoni leaders had been executed. The execution was carried out days before the appeal period set by military decree was to expire. This cruel and unjust act confirmed the rigged nature of the entire affair.

The world and environmental justice activists in the grassroots networks, such as the Friends of the Earth International, were numbed by this blood-letting; November 10 has become a day of remembrance of the martyrs of resistance against mining, oil and gas. The lesson that this sends is a confirmation that killing the messenger merely throws up more messengers. The message cannot be killed.

Sister McCarron watered my activist roots through books and other resources that she donated to the then very young Environmental Rights Action. The books all showed strong links between social justice and human dignity. They all stressed that the heavenly focus is not genuine if it negates our humanity and our need to be concerned with the state of the planet and the people. It is no surprise that she forged a strong connection with Ken Saro-Wiwa and was a strong supporter of the Ogoni and wider Niger Delta struggles.

Meeting Sister McCarron in Leitrim, Ireland at a campaign stop against fracking in August 2012, brought a new spark of inspiration to me, seeing that her commitment to securing and preserving a sane environment is a lifelong affair. The campaign in County Leitrim was against fracking in Ireland. From there I moved to Erris in County Mayo on a solidarity visit with the people organised by Friends of the Earth, Ireland. Sister McCarron has worked with the people in this county, supporting their struggles against the building of gas pipelines and a refinery in the area by Shell. That struggle, epitomised by the existence of the Rossport Solidarity Camp and the nine white crosses (for the Ogoni nine) opposite the entrance to the refinery at Ballinaboy, has gone on for over a decade, in the face of intimidation, beatings and jailings.

In the poem, *For Sr. Majella McCarron*, written in June 1995, we read the pointed question and answer:

> What is it, I often ask, unites

County Fermanagh and Ogoni?
Ah, well, it must be the agony
The hunger for justice and peace
Which married our memories
To a journey of faith.

Ken Saro-Wiwa was a man of peace. As these letters clearly show, his struggles were built on the platform of non-violence through the mobilisation of the mass of Ogoni people using socio-cultural and educational tools. His disdain for violence can be seen in some of his writings, especially in the novel, *Sozaboy*.

He was a prolific writer, and is even said to have engaged in writing anonymous letters to editors of newspapers while he was in the secondary school. His books such as *On a Darkling Plain* focused on his Biafra-Nigeria civil war experiences and ruffled the feathers of those who do not agree with his interpretation of the events. Saro-Wiwa's works included children's books as well as the highly successful satirical television series, *Basi & Co*. One of his short stories, *Africa Kills Her Sun*, first published in 1989, foreshadowed his execution and denial of a decent grave. The book, *The Politics of Bones*, written by J. Timothy Hunt about Owens Wiwa, Saro-Wiwa's brother and comrade in the struggle, fittingly captures the struggle for a decent resting place for Saro-Wiwa.

Saro-Wiwa captured the hearts of many through his writings and showed that enduring change required fundamental reorientations to ensure that his people worked for the good of the collective rather than for self-aggrandizement.

Perhaps the letters set out in this book provide a partial completion of his diaries from an earlier period of detention, *A Month and a Day*. The final part may not be known until we know the contents of the letters that Sister Majella wrote in reply to these.

Perhaps one of the most critical contributions of Ken Saro-Wiwa to the struggle for the salvaging of the Ogoni environment and by extension the larger Nigerian environment was his participation in the production of the Ogoni Bill of Rights (1990) by MOSOP. That bill of rights remains the cardinal articulation of the demands for holistic justice for the Ogoni people. The Bill pioneered the way for the formulation of bills of rights by other ethnic nations in Nigeria, including the Kaiama Declaration by Ijaw youths (1998), the Oron Bill of Rights (1999), the Aklaka Declaration (1999) of the Urhobo Economic Summit and others.

We do not exaggerate if we say that Ken Saro-Wiwa was prescient, even prophetic, in his writings. A recent assessment of the Ogoni environment by the United Nations Environment Programme (UNEP) clearly vindicates the complaints of the Ogoni that their environment had literally been killed by the polluting activities of the international oil companies led by Shell, as well as by their Nigerian counterpart, the Nigerian National Petroleum Corporation. According to UNEP, it would take about 30 years of focussed and robust clean up efforts to restore the Ogoni environment. A year and a half after the publication of the UNEP report, the Ogoni people and all watchers of the territory are scandalised that the Nigerian government has not declared a state of environmental emergency in the area.

The chapters prefacing the letters and poems provide excellent analyses that help us understand the person of Ken Saro-Wiwa as a selfless, dogged fighter, who did not project himself as a hero but rather fought for the collective Ogoni nation. Laurence Cox's chapter on "Ken Saro-Wiwa in Political Context: Social Movements in the Niger Delta" brings up the interesting discourse on "the curse of oil." He shows how the struggle against Shell in County Mayo, Ireland, and the use of security forces to counter dissent draws strong parallels between Ireland and Nigeria.

Cox also calls for caution when Norway is presented as an exception to the rule of oil not being of benefit to local populations in countries where it is extracted. The vital point here is that Norway was different at the onset of oil activities there, not by the benevolence or brilliance of government or oil sector players but by the efforts of popular movements that were determined to ensure that oil wealth benefited all. Analysts also show that a strong workers movement ensured that technology was bent to ensure safety rather than merely profits. As Norwegian oil begins to peter out, the country's oil sector players are seeking to drill in fragile ecosystems, including in the Arctic regions—a possible harbinger of fracturing of the image that the Norwegian nation has built and that has lured nations across Africa to think that the Ogoni experience cannot be replicated in their territories. The thinking of nations that change can come about without robust resistance is nothing but a pipedream.

The last speech that Ken Saro-Wiwa made, before the rigged tribunal that condemned him, reads like a letter to the world; his challenge to everyone remains valid. Here is what he said:

I repeat that we all stand before history. I and my colleagues are not the only ones on trial. Shell is here on trial and it is as well that it is represented by counsel said to be holding a watching brief. The Company has, indeed, ducked this particular trial, but its day will surely come and the lessons learnt here may prove useful to it, for there is no doubt in my mind that the ecological war that the Company has waged in the Delta will be called to question sooner than later and the crimes of that war be duly punished. The crime of the Company's dirty wars against the Ogoni people will also be punished.

On trial also is the Nigerian nation, its present rulers and those who assist them. Any nation which can do to the weak and disadvantaged what the Nigerian nation has done to the Ogoni loses a claim to independence and to freedom from outside influence. I am not one of those who shy away from protesting injustice and oppression, arguing that they are expected in a military regime. The military do not act alone. They are supported by a gaggle of politicians, lawyers, academics and businessmen, all of them hiding under the claim that they are only doing their duty, men and women too afraid to wash their pants of urine.[4]

This collection is a unique gift, a treasure to humanity. The letters provide strands that can be woven into manuals for popular struggles. In places, their humour is quite hilarious. And revealing. Happily, rather than being moved to despair, they will help the reader make up his or her mind on whether to raise a voice or stay silent in the face of tyranny. This publication is a fitting tribute to this gentle but strong leader, and it is hoped that it will help clarify his enigmatic personality and inspire positive action against oppression and ecocide anywhere and everywhere. Our struggles may be in different locations, but they are one and the same. We are all Ogoni. And in the words of Ken Saro-Wiwa, "the Ogoni story will be told." We are happy to say that this book tells a critical part of that story.

4. "Ken Saro-Wiwa's Final Address to the Military-Appointed Tribunal." *Earth Island Journal.* 11.1 (Winter 1995): 25

Introduction to the second edition

HELEN FALLON, ÍDE CORLEY & LAURENCE COX

MUCH HAS HAPPENED since Sr. Majella McCarron (OLA) donated her collection of material relating to Ken Saro-Wiwa, including letters, poems, artefacts and recordings, to Maynooth University in November 2011. Work on the conservation, preservation and cataloguing of the archive has been completed and the digitised letters are now available on open access. Links to the letter and other materials, including photographs, YouTube videos, articles, blog posts and records of various events, can be accessed online via the Maynooth University Library Ken Saro-Wiwa Library Guide.

In 2013, we had the privilege of welcoming Ken's brother, Dr Owens Wiwa, to the university to launch the first edition of *Silence Would be Treason: Last Writings of Ken Saro-Wiwa*. During his visit, he viewed the archive in the Library Special Collections Reading Room and was deeply moved to read the letters and see the MOSOP cap his brother had worn.

In November 2015, to mark the 20th anniversary of Ken Saro-Wiwa's untimely death, the university also hosted a visit by his daughter, the renowned travel writer, Noo Saro-Wiwa. While at Maynooth, she presented a reading from her award-winning book, *Looking for Transwonderland: Travels in Nigeria* to a large audience and was interviewed for the Ken Saro-Wiwa Audio Archive, an innovative open access sound archive which is described in a new chapter in this second edition. Noo has generously contributed a foreword to the second edition.

Funds from the sales of the earlier edition of this book were used in the establishment of a research award in Ken Saro-Wiwa's name. On the 10th November 2015, Noo presented the inaugural Maynooth University Ken Saro-Wiwa Postgraduate Award to history PhD candidate Graham Kay. His PhD thesis explores the role of oil in Anglo-German geopolitical competition in the lead-up to the First World War – a context which has continued to shape Nigerian and Ogoni history. His research is outlined in a new chapter in this second edition.

That history has not stood still. Since the first edition of these letters were published, agreement has been reached with Shell and other oil companies on a five-year programme cleaning up areas in Ogoniland devastated by oil spills, with controversy raging as to why the programme has not yet (2017) taken place. In Ireland, Shell's pipeline in Mayo has been completed, but popular pressure has forced a legal ban on onshore fracking. In the wider world, the political and economic centrality of oil has come under question as never before, with declining prices, the growth of renewables and increasing recognition of the reality of climate change. The role of the Ogoni struggle in this historical transition makes these letters an important document far beyond the borders of Rivers State or Nigeria itself.

In November 2017, Mark Dummett from Amnesty International (London) gave the keynote address at the Maynooth University Ken Saro-Wiwa Seminar. He outlined the key findings of Amnesty International's 2017 report on Shell's activity in the Niger Delta.[1] Mark's afterword to this volume gives an insight into the current situation in Ogoni.

It was with great sadness we learned of the death of Ken Saro-Wiwa's son, Ken Wiwa, in October 2016. Maynooth University is indebted to Ken Jr. for his support for both this book and our other endeavours to highlight the issues his father and eight colleagues (the Ogoni Nine) campaigned and died for.

In the same month as Ken Jr.'s sudden death, the first international exhibition of material from the collection opened at Quinnipiac University in Connecticut. This was one of many initiatives to create awareness of the collection and the issues it embodies. Similarly, at Maynooth University, items from the collection have been regularly exhibited to mark key events and coincide with conferences and seminars with an environmental or conflict resolution theme. A local public library selected *Silence Would be Treason: Last Writings of Ken Saro-Wiwa* as their reading club book and children from nearby schools have visited the library, viewed the letters and discussed the issues surrounding the conflict over natural resources in class. In July 2016, 25 schoolchildren with Nigerian backgrounds visited the Maynooth Special Collections to learn about Ken Saro-Wiwa and the issue of environmental rights. The children were intrigued to learn that his letters to Sister Majella were smuggled out of military detention in breadbaskets.

1. Amnesty International (2017). *A Criminal Enterprise? Shell's Involvement in Human Rights Violations in Nigeria in the 1990s. https://www.amnesty.org/download/Documents/AFR4473932017ENGLISH.PDF*

Commenting on the event, Sister Majella remarked:

> I wanted the archive to be available to activists and researchers. This the Library is doing very well. Making it available, in an accessible way, to the schoolchildren today, is another great way of ensuring the Ogoni story is told; and it was all great fun too!

As the second edition goes to press, plans are underway for a Ken Saro-Wiwa travelling exhibition in Ireland. The university library is working with local schools to run a competition for a poem that embodies the ideals of Ken Saro-Wiwa and a filmmaker is actively exploring using the book in a forthcoming film about the life of Ken Saro-Wiwa.

In a letter dated 1st December 1993, Ken Saro-Wiwa advised Majella McCarron:

> Keep putting your thoughts on paper. Who knows how we can use them in future. The Ogoni story will have to be told! [2]

In this second edition of *Silence Would be Treason: Last Writings of Ken Saro-Wiwa*, we continue to tell the Ogoni story and publicise the Ogoni cause. We are particularly pleased that this volume will also be available on open access and thank our publisher, Firoze Manji and Daraja Press for facilitating this. Maynooth University is particularly indebted to Firoze for his valued contribution and ongoing dedication to this project.

2. *Maynooth University Ken Saro-Wiwa Archive PP7/2*

INTRODUCTORY

Ogoni flag signed by Ken Saro-Wiwa in 1994. Maynooth University Ken Saro-Wiwa Archive.

The Ken Saro-Wiwa Collection at the Library, Maynooth University, Ireland

HELEN FALLON

SR. MAJELLA MCCARRON (OLA) PRESENTED, on 10th November 2011, a collection of personal correspondence and 27 poems she had received from Nigerian writer and activist Ken Saro-Wiwa to the Library of Maynooth University. The date marked the sixteenth anniversary of Saro-Wiwa's execution. Professor Philip Nolan, President of Maynooth University, accepted the gift from Sr. Majella on behalf of the University, saying, "the collection cast a very human eye on what was one of the late 20th century's most troubling geopolitical issues".

The collection comprises 28 letters to Sr. Majella, 27 poems, seven video cassette recordings of visits and meetings with family and friends after Saro-Wiwa's death, a collection of photographs and other documents, including articles, reviews, flyers and maps relating to Saro-Wiwa's work and the work of Sr. Majella on the cause of the Ogoni people, both in Nigeria and Ireland.

Letters from Ken Saro-Wiwa Archives. Maynooth University Ken Saro-Wiwa Archive.

The letters were mainly handwritten between 20th October 1993 and 14th September 1995. The two earliest, dating from 1993, are from the period before Saro-Wiwa was taken into military detention: the first having been sent from his office in Port Harcourt, the second from Lagos. In May 1994 Saro-Wiwa and several other activists were placed in military detention in Port Harcourt. The letters written between July 1994 and his execution in November 1995 were smuggled out of detention in food baskets.

The letters cast light on Saro-Wiwa as a political activist, a writer, a family man and a personal friend to Sr. Majella. He wrote other letters during his captivity; apart from those to family and friends, he corresponded with Ethel Kennedy, Anita Roddick, Nelson Mandela, novelist William Boyd and other international figures. What makes the collection of letters to Sr. Majella unique is the relatively large number (28) and the combination of both personal and political content. Saro-Wiwa's letters to Sr. Majella paint a picture of a well-educated and articulate Nigerian writer and illuminate his efforts to help the Ogoni people, who number less than 500,000 and live in the south eastern part of the Niger Delta.

Maynooth University Ken Saro-Wiwa Archive.

While the letters tell a story of Ken Saro-Wiwa, we do not have the letters Sr. Majella sent to him to gain a more complete picture of this nun who left her rural home in Derrylin, County Fermanagh, to join the Missionary Institute of

Our Lady of Apostles (OLA) in Cork in 1956. From there, armed with a science degree from University College Cork, she travelled to Nigeria. She initially taught science in a secondary school. Later, she completed postgraduate studies both in the University of Lagos and the University of Ibadan. In 1990, she was asked by the Missionary Institute of Our Lady of Apostles to do some work in Nigeria on its behalf. The Institute, a member of the Brussels-based Africa-Europe Faith and Justice Network (AEFJN), which lobbies the European Union on behalf of communities adversely impacted by European business, asked Sr. Majella to identify such communities in Nigeria. AEFJN offered to speak on their behalf at the EU. The oil problem in the Niger Delta region of South Eastern Nigeria was severe at the time, with major environmental damage being wrought by oil extraction. Saro-Wiwa, the leader of the Movement for the Survival of the Ogoni People (MOSOP), was organising a non-violent campaign against the environmental destruction of the Ogoni area of the Niger Delta. Sr. Majella brought the Ogoni complaint of environmental destruction by Royal Dutch Shell to the attention of AEFJN. He would eventually visit their office to offer his appreciation. Today AEFJN continues to monitor oil and gas impact throughout Africa.

When Sr. Majella selected this project, she had no idea how big it would become or of the strong relationship she would develop with Saro-Wiwa. She was initially introduced to him by Lynn Chukura, a prominent member of the Association of Nigerian Authors of which Saro-Wiwa was once president. Now deceased, Chukura, who was from Philadelphia, had lectured at the University of Lagos, the University of Ibadan and the University of Legon, Ghana. She was to give shelter to the wife and child of Ken's brother Owens Wiwa in Lagos while he was avoiding arrest.

Sr. Majella was lecturing in Education at the University of Lagos when she met Saro-Wiwa. Academic institutions in Nigeria had frequent strikes and during one such period, she researched Ogoni issues. She visited Saro-Wiwa's office in Lagos to consider with him how best to advance the Ogoni concerns. When Ogoni villages were destroyed in September 1993, Sr. Majella worked with the Daughters of Charity in Port Harcourt and Trócaire in Dublin to obtain a European Community grant for village relief. The first letter in the collection is Saro-Wiwa's grateful response to her and offers MOSOP's goodwill for the Church administration of this grant.

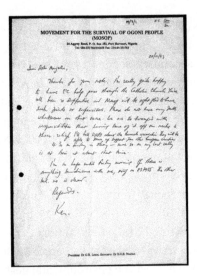

Letter from Ken Saro-Wiwa to Sr.
Majella McCarron, 20/10/93.
Maynooth University Ken Saro-Wiwa
Archive.

The collection contains photographs taken by Sr. Majella recording the destruction of the Ogoni villages, an act that was locally explained as a reactive punishment by unexplained forces to the campaign undertaken by MOSOP.

Ogoni villagers survey destroyed canoes at Kaa, 5th January 1994. Photograph by Sr. Majella McCarron. Maynooth University Ken Saro-Wiwa Archive.

Destroyed buildings at Kaa Market in Ogoniland, 1994. Photograph by Mary Sweeney. Maynooth University Ken Saro-Wiwa Archive.

Some Ogoni inhabitants survey the impact of violence along the waterfront at Port Harcourt, 5th January 1994. Photograph by Sr. Majella McCarron. Maynooth University Ken Saro-Wiwa Archive.

Another set of photographs records the efforts of Trócaire and the Daughters of Charity in the Catholic diocese of Port Harcourt to reconstruct houses in ten of these villages by means of the EC emergency aid she had helped to secure.

Relief workers rebuild Ogoni villages with the EC emergencu funding that was released to the Diocese of Port Harcourt through Trócaire in Ireland. Maynooth University Ken Saro-Wiwa Archive.

The collection also includes a third set of photographs which she received from a refugee camp set up in the Republic of Benin for fleeing Ogoni people. She was to keep in contact with this group and many other refugees in Europe and the US for several years.

Ogoni Refugee Camp in the Republic of Benin, 1997. Maynooth University Ken Saro-Wiwa Archive.

Towards the end of 1993, Ken Saro-Wiwa discussed his concerns about developing the leadership potential of MOSOP with Sr. Majella. From her lecturing experience, she was familiar with a programme which used the psychosocial method of Paolo Freire, a Brazilian adult educator. His approach had spread through many grassroots communities internationally. In Nigeria, it had helped to shape the Development Education and Leadership Service administered by a team within the Catholic Church. Sr. Majella sought the assistance of one of its members, St. Patrick's Society priest Tommy Hayden, and arranged a planning meeting with Saro-Wiwa in early October 1993. The first of a five-phase programme took place in 1993. While a second phase was planned for 1994, this was cancelled amid an outbreak of hostilities, turmoil and destruction.

It was an extremely dangerous time in Nigeria, with the Biafran war (1967-70) and the expulsion of missionaries still very prominent in people's minds. The special security forces were very active and to be seen visiting the office of Saro-Wiwa was dangerous for Sr. Majella and others. "Anyone protesting in a military dictatorship places themselves in a dangerous position," Sr. Majella commented in conversation with the author. She wrote observations on the Ogoni Movement, while Saro-Wiwa gave her regular updates, seeing her as a contact with the outside world–a way of getting his ideas and ideals out to a wider audience. "Ken and I used to talk at length about the problems concerning the Ogoni people. He appreciated my analysis and thoughts. I think he felt I was a benign, spiritual presence. I was a trusted witness," Sr. Majella remarked.

In May 1994, Saro-Wiwa and other members of MOSOP were arrested. The twenty six letters dating from this time until his death were written in military detention. In a letter of 13th July 1994, he tells Sr. Majella he has managed to get a computer smuggled in so that he could continue with his writing. The urge to write is not just tied to the ideological conviction of the writer; for centuries, the commitment to write, despite expense and logistical difficulties, speaks of the pressing need for human contact. From military detention, Saro-Wiwa shared with Sr. Majella his thoughts on the Ogoni cause, his love for his family and friends and his passion for writing. On 13th July 1994, he wrote:

> Have you been able to go through my essays: I wonder if they can be made into a book or should I take just the central ones and put them alongside the newspaper articles which I have collected? Seems to me that would be a better idea.[1]

In August 1994, Sr. Majella returned to Ireland for a sabbatical, having decided not to renew her contract at the University of Lagos where she had taught for 13 years. On 30th July, prior to her departure, Ken writes simply,

> "I will miss you in the year you are going to be away!"[footnote]MU archive PP/7/6

The conversations that had begun in the Lagos office continued on paper; the high esteem in which Saro-Wiwa held Sr. Majella, and the value he placed on their contact as a window on the outside world, shines through in his letters. On 15th August 1994, he wrote to her in Ireland: "A million thanks for your letters. They are so entertaining, so encouraging, & they give me those intimate details of my family which no one else gives. God bless you & keep

---

1. MU archive PP/7/3[/f ootnote]

you for us!"[2] There is poignancy in his reference, in a letter dated 16th September 1994, to Sr. Majella's visit to his family in London and her introduction to his newborn son: "It's hard to think you've been away for a whole month! How fast does time fly? I saw your picture holding my young son Kwame whom I'm still to see."[3] A fortnight later, on 1st October, he writes again, stressing her role in helping the Ogoni people:

> I long to see you back in Nigeria, helping, among others, to guide the Ogoni people through the wilderness. You don't know what help you have been to us and to me personally, intellectually. God grant that you do return to us. I'm counting the days.[4]

Sr. Majella did in fact return to Nigeria, on a short visit at the request of Trócaire, to finalise paperwork relating to various grants she had secured. She was not allowed to visit Saro-Wiwa during this period, but continued to correspond with him. The value of the correspondence in breaking his isolation was noted by Saro-Wiwa in a letter of the 24th October:

It was, as usual, most informative and made me miss you more than ever. You are the only one who gives me an almost complete picture. Had you gone to Ogoni, I'd have got a complete picture—local, national and international. It can be terrible here when one cannot phone or discuss with others.[5]

In October 1994, Saro-Wiwa was awarded the Right Livelihood Award, a Swedish prize presented annually, to honour people who make a significant contribution to solving modern-day challenges. He wrote his acceptance speech from his detention centre and Sr. Majella reviewed it for him. His request to attend the award ceremony, recorded in a letter of the 29th October, was refused by the military authorities. In December, Sr. Majella travelled to Stockholm for the ceremony. Photographs from the event are part of the collection donated to the Library. There is also an Ogoni flag which Saro-Wiwa listed among items for Majella to take to include in an exhibition in Stockholm to coincide with the presentation of the award. On Christmas Eve 1994, Saro-Wiwa wrote to her:

2. MU archive PP/7/7
3. MU archive PP/7/8
4. MU archive PP/7/9
5. MU archive PP/7/12

Dear Sr. M,

Thanks a million for your diaries of August, November and early December. It was really wonderful to hear all about the Stockholm Award Ceremony from your careful observant pen.[6]

Sr. Majella devoted her sabbatical year to saving the lives of the Ogoni Nine through campaign work in Ireland. This led to the establishment of Ogoni Solidarity Ireland. Trócaire gave her the use of a phone and helped her to publicise the case of Saro-Wiwa and the eight others who were to be tried by a military tribunal. Photographs are included in the collection from this campaign. Her efforts offered some comfort to Saro-Wiwa, who wrote on the 21st March 1995:

> I have seen your work & your pictures in the *Irish Times* and I think you yourself might be surprised how far those Ogoni bells are ringing now, and how you yourself have become the bellman. I thank God for your presence among us.[7]

In the last letter before his execution, dated 14th September 1995, he wrote of her letters:

> [...]I believe that I've got everything you have sent thus far. Some of them come rather later and out of sequence, but I do get them. Because I keep them around me just to read and re-read them, I've had two of them seized lately. I hope that I will get them back, anyway, some day.
>
> I expect that you have now started your new assignment and am really happy for you. It is hard to think that you will no longer be with us here in Nigeria, but it may well be that we shall be better served by your being away.[8]

She received this, his final letter, on 10th October, from his son, also named Ken, when he arrived in Belfast to receive a Nobel Peace Prize nomination secured by Mairéad Corrigan of the Peace People and supported by Amnesty

---

6. MU archive PP/7/20
7. MU archive PP/7/23
8. MU archive PP/7/28

International Northern Ireland, Trócaire Northern Ireland and the Bodyshop. The nomination was planned for Saro-Wiwa's fifty-third birthday and was intended as a public appeal for the lives of the Ogoni Nine. One month later, their lives were taken at the behest of a military tribunal. It is very fitting that the collection now contains an e-mail from Mairéad Corrigan, herself a Nobel laureate, sent to Sr. Majella on reading newspaper coverage of the donation to Maynooth:

> Dear Majella,
>
> This is wonderful news that the life and writings of Ken Saro-Wiwa will be held at National University of Ireland Maynooth [9]. His life is a great testimony to the human spirit of love and compassion and self-sacrifice for others.
>
> I had the great honour of having breakfast with Ken Saro-Wiwa during a Conference of UNPO in The Hague some years ago. He spoke with such love and passion about the plight of his people and also with love of his family particularly his son, who then was working as a journalist in London.
>
> Ken would be pleased that you, his dear and faithful friend, have placed this material in the National University of Ireland Maynooth.
>
> It's amazing how things like letters remind us of our own connections and I am reminded too that you wrote to the Peace People about Ken and his Movement, so thank you Majella for all your work in supporting, during his lifetime, Ken and his people.

Ken Saro-Wiwa is considered to be one of the great environmental activists of the late 20th century and his letters reflect his passion for peace and justice around the world. Sr. Majella—an Irish female missionary charged with keeping Saro-Wiwa briefed and acting as the carrier of his message globally—is vital to this correspondence, having called the letters into existence. In his captivity, Saro-Wiwa must have drawn comfort from knowing the lengths to which Sr. Majella would go to ensure the delivery and protection of his written words. At a time when communications technology decreases the habit of traditional letter-writing, the enduring value of the letter in disseminating the voice of the oppressed remains. As a well-established writer,

9. since renamed Maynooth University – Ed.

it is likely that Saro-Wiwa knew the letters were going to be read and considered in different contexts.

In gifting these letters to the Library at Maynooth University, Sr. Majella is ensuring the Ogoni story will continue to be told in many different contexts. These letters will be studied by students and researchers from a wide range of disciplines—Anthropology, Education, History, Economics, Geography, Sociology, Politics, Development Studies, African Studies, Archival Studies and Literature—all of which are represented in the courses offered at Maynooth University.

For this reason, the editors of this volume are grateful to Maynooth University sociology student, John O'Shea, who created the initial link between Sr. Majella and the University Library. In 2010, O'Shea interviewed her while working on his MA thesis. She told him about the Saro-Wiwa material she held and expressed an interest in finding an appropriate home for it, knowing the value this collection would have to present and future generations of scholars and activists. He contacted the Library and we immediately set about acquiring this unique collection. The Library already holds significant archives of writers, social and political figures, and missionaries and this collection will prove a most valuable addition in this area. The Ken Saro-Wiwa Collection complements our extant African holdings which comprise, among other works, a major collection of bibles representing 298 African languages, including Khana, Saro-Wiwa's native tongue. Both on its own and in the context of broader themes which it so deftly illustrates, this donation marks a major new addition to our collections.

It is very timely that this donation has come shortly after the establishment of the Edward M Kennedy Institute for Conflict Intervention at Maynooth University, specialising in research on conflict resolution, leadership and civic engagement. The Saro-Wiwa correspondence will bring the lessons of the need for open dialogue around conflict to a wider audience—something the author was prevented from doing by his execution. Research of this nature needs to be happening at university level, where there are structures to support it and opportunities to share the findings globally. It is particularly appropriate that this research should be carried out in Maynooth University which has a long-standing commitment to equality and social justice.

The collection is now available in the Maynooth University Library for consultation. Archival and conservation work has been carried out, including cleaning and repairing documents that require attention, cataloguing each item and transferring the letters to acid free packaging materials. Future exhibitions will offer the opportunity for people to view the collection and the Library will provide a neutral space for people to become actively involved with Saro-Wiwa's thoughts and experiences through his letters and other writings. Journal articles, research papers, theses and other work carried out on the Saro-Wiwa archive by staff and students of Maynooth University will be available—not just to the international research community but also to those involved in social movements all over the world. The letters have now been digitised and are available on open access.

The videos, which form part of the gift, record Sr. Majella's visits to and meetings with the family and with the traditional authorities in Ken Saro-Wiwa's home village of Bane as well as with members of the Methodist Church. There is an interesting Irish link here in that this was the earliest Christian missionary church in Ogoni and its missionaries were from the Kingston family of Cork, Ireland. This missionary family coordinated the translation of the Bible into Khana. The videos have been converted into more modern media to allow continuing access to the contents.

After the November 1995 executions of the Ogoni Nine, Sr. Majella continued to campaign on behalf of twenty more Ogoni detainees who were freed six years later. Much of the publicity material from this period is now with the collection.

Dublin protest. Maynooth University Ken Saro-Wiwa Archive.

Gabriel Byrne and others at AFri-Famine Walk, NW Ireland. Maynooth University Ken Saro-Wiwa Archive.

Majella McCarron at Shell-to-Sea protest in Ireland. Maynooth University Ken Saro-Wiwa Archive.

Ogoni Solidarity Ireland protest in Dublin. Maynooth University Ken Saro-Wiwa Archive

Protest outside Nigerian Embassy, Dublin. Maynooth University Ken Saro-Wiwa Archive.

These efforts were supported in Ireland by Trócaire, Action from Ireland, Amnesty International, the Bodyshop, the Peace People, many local groups from Cork to Donegal, a school in Dunboyne, and Belvedere College in Dublin. Ogoni Solidarity Ireland provided the essential focal point and Saro-Wiwa's letters to Sr. Majella from May 1994 to July 1995 informed the Irish campaign before, during and long after the deaths of the Ogoni Nine. The material has been a source for the annual Saro-Wiwa Memorial Seminar organised by Sr. Majella in different venues every year since 1995, to consider the impact of multinational business on local communities in Ireland and elsewhere. Photographs from the collection, particularly that of a large mural remembering Ken Saro-Wiwa and his eight companions erected on the tenth anniversary of their death, record the connection between Ogoni and Erris in Mayo, Ireland. There are also photographic records of nine white crosses, situated opposite the Shell terminal at Ballinaboy, which were carried in campaigns. The letters informed Irish engagement with Shell in Nigeria and Ireland.

The letters, videos and other items are important historical and cultural artefacts, coming out of specific historical circumstances and specific human activity. The collection is part of the heritage of Saro-Wiwa, part of the heritage of Nigeria and particularly the Ogoni, and part of the heritage of the Sisters of Our Lady of the Apostles as represented by Sr. Majella in Nigeria. This wonderful donation comes at a particularly appropriate time. Our major new library extension has a purpose-built Archives and Special Collections area where material can be maintained, exhibited and consulted in the best possible environment.

In a letter of the 1st December 1993, Saro-Wiwa urged Sr. Majella: "Keep putting your thoughts on paper. Who knows how we can use them in future. The Ogoni story will have to be told!"[10] In gifting this material to the Library, Sr. Majella is ensuring that the Ogoni story will be told.

10. MU archive PP7/2

"Ogoni Story will have to be told"- Ken Saro-Wiwa. Maynooth University Ken Saro-Wiwa Archive.

Ken Saro-Wiwa and West African Literature: The Politics of Language

ÍDE CORLEY

IN A LETTER WRITTEN from his detention cell in Port Harcourt in July 1994, part of which now forms the Preface to his memoir, A Month and A Day (1995), Ken Saro-Wiwa regretted that he did not have time to include an account of what had happened in Ogoniland[1] between his release, after his first detention, on 22nd July 1993–when A Month and A Day ends–and his re-arrest on 22nd May 1994: "[T]he events of that period would have enriched this book even more".[2] The 28 letters in this volume addressed to Sr. Majella McCarron date from 20th October 1993, roughly three months from the time that A Month and a Day ends, to 14th September 1995, just under two months from the date he was executed. They were accompanied by 27 poems which he enclosed over the course of the correspondence. One of the poems dedicated personally to McCarron and two of the letters, including the very last–which, as Helen Fallon has noted here, is the author's last recorded letter–have previously appeared in an addendum to the second edition of A Month and a Day (published as A Month and a Day and Letters) in 2005. This volume places them into the broader context of the two-year correspondence, giving us a better picture of their extraordinary friendship and shared concerns. It also sheds new light on the chaotic period 1993-1995, which the author sadly perceived that he might not live to relate.

The narrative generated by these documents extends the published record of Saro-Wiwa's life but it is necessarily partial and shaped to a certain extent,

---

1. Saro-Wiwa preferred to use the term "Ogoni" in preference to "Ogoniland" because, as he wrote in A Month and a Day "to the Ogoni, the land and the people are one and are expressed as such in our local languages". See Ken Saro-Wiwa, A Month and a Day and Letters, Foreword by Wole Soyinka, (Oxford: Ayebia Clark Publishing Limited, 2005) p.3. I use the term "Ogoniland" here to clarify issues for readers who are unfamiliar with the Nigerian contexts of his writing.
2. Ibid. p.2

as can only be expected, by the disposition, interests and activities of the addressee. It reveals his ongoing participation in and reactions to contemporary events alongside his plans for the future but, as a series of letters, does not provide the kind of supporting ethnographic and historical information that characterize his more formal political statements in texts such as *Genocide in Nigeria* (1992) or *A Month and a Day*. Saro-Wiwa has often been described as a contradictory figure: an Ogoni ethno-nationalist who upheld federalism during the Nigerian Civil War, 1967-1970; a democrat who appeared, at times, to have embarked upon "a drive for personal authority"[3]; a proponent of non-violent protest whose execution was arranged on charges of incitement to murder. Many of these apparent contradictions stem from the efforts of his enemies to discredit him for their own private gains and circulate in the context of a poor general understanding of his complex, interstitial positioning as a representative of an ethnic minority group in postcolonial Nigeria. In this short essay, I aim to situate his work within a longer tradition of Nigerian political thought and writing. I characterize Saro-Wiwa as a "pragmatic nationalist"[4] whose actions were ultimately informed by his "desire to live in a just society"[5] rather than by any deeply held belief in the intrinsic identity or separate destiny of the Ogoni people.

In West Africa, both at the time Saro-Wiwa was writing and earlier, literary activity had been crucial to the formation and articulation of oppositional projects and politics.[6] Although indigenous, oral cultures remained relevant and vibrant during the colonial era, the anti-colonial nationalist movements which arose alongside, and in protest against, colonial regimes in the late nineteenth to early twentieth century heavily depended upon the establishment of African-owned, europhone newspapers and printing presses.[7]21 The importance of europhone print cultures to anti-colonial nationalisms in West Africa related, in part—as any student of West African history well

---

3. See Ato Quayson, *Calibrations: Reading for the Social* (Minneapolis: University of Minnesota Press, 2003) p. 70
4. See Tommie Shelby, *We Who Are Dark: The Philosophical Foundations of Black Solidarity*, (Cambridge, Massachusetts and London, England: The Belknap Press of Harvard University Press, 2005) p. 28
5. Ibid. p. 28
6. See Stephanie Newell, *West African Literatures: Ways of Reading*, Oxford Studies in Postcolonial Literatures, (Oxford: Oxford University Press, 2006) especially pp. 19-20
7. Ibid. p. 19

knows—to the colonial re-distribution of territory on the continent by European powers at the Berlin Conference in 1884-5. At this conference, the modern political boundaries of African states were drawn up by European technocrats; they were determined not, as in Europe, by the shared identities, languages and cultures of indigenous social groups but by the commercial interests and needs of European empires and their desire to prevent imperial conflicts over strategic territories and resources.[8] In practical terms, this meant that indigenous populations who shared common languages, traditions and cultural practices were often divided among colonial states which imposed different and distinct languages, systems of administration—and, effectively—histories upon them.

As the most populous colonial state in Africa, the Colony and Protectorate of Nigeria (established in 1914) contained the most heterogeneous and diverse population. While home to three majority language groups—the Hausa-Fulani of the North, the Yoruba of the South-West and the Ibo of the South-East—collectively the populations of the territory speak over 250 distinct languages. During and after the colonial era, as Saro-Wiwa's writing attests, many Nigerian language groups recognized and maintained pre-existing local and translocal ties with kin, locality and ethnic nation over and above any obligations to the state. The colonial project was further complicated by the rise of new vernaculars among urban migrants as they creatively adapted to new and ever more severe conditions of survival.

Within this fractured social environment, a minority group of Western-educated indigenous elites began to use English language in the early twentieth century both to challenge colonial authorities and to negotiate ethnic conflicts and differences. Because decolonization was never simply a domestic issue, the Anglophone leaders of multi-ethnic Pan-African nationalist movements in Nigeria creatively engaged with the leaders of other Pan-African nationalist groups in neighbouring countries in West Africa, as well as in East and Southern Africa, Britain, France, the Caribbean and North America, to promote the project of decolonization globally.[9] Modern Pan-African political movements aimed to promote unity among black social

8. See Thomas Pakenham, *The Scramble for Africa: White Man's Conquest of the Dark Continent 1876-1912* (New York: Avon Books, 1991) pp. 239-256
9. Vincent Bakpetu Thompson, *Africa and Unity: The Evolution of Pan-Africanism* (London and Harlow: Longmans, 1969); Immanuel Wallerstein, *Africa: The Politics of Unity*

groups and to project an ideal of African unity to an international audience. In doing so, they frequently diminished or underplayed the historical tensions and conflicts among indigenous African groups.[10]

Nigeria was one of the most resource-rich colonies and the political situation there was complicated further by the federal structure designed by the exiting colonial administration to secure British commercial interests after independence. Once the Atlantic Charter—which upheld the right of all people globally to self-determination—was recognized after World War II, the British Parliament quickly moved in 1946 to approve the Richard's Constitution for the territory with a view to blocking the evolution of multiple national polities in favour of a centralized administration. The Richard's Constitution helped establish a federation of artificial regions— Northern, Western, and Eastern—which, as Saro-Wiwa repeatedly complained, generated, within each region, "a dominant and dominating ethnic group"[11] to which a number of smaller ethnic nations were politically subjugated.[12] It gave rise to the troubling contradictions which later came to characterize the postcolonial state; in Saro-Wiwa's words, "there was 'unitarism' at regional level and 'federalism' at the centre".[13]

Saro-Wiwa was an Ogoni, one among the many smaller ethnic nations (sometimes described as "micro-minorities"[14])—also including, for example, the Annang, Efik, Ekpeye, Ijaw, Ikwerre, Etche, Odual, Sankwala, Ukele,

(London: Pall Mall Press, 1967); Robert Young, *Postcolonialism: An Historical Introduction* (Oxford, England and Maldon, Massachusetts: Wiley-Blackwell, 2001) pp. 217-274

10. Ato Quayson, *Calibrations: Reading for the Social* (Minneapolis: University of Minnesota Press, 2003), pp. 30-55; Eileen Julien, "The Extroverted African Novel" in Vincent Moretti Ed. *The Novel*, Volume 1, History, Geography and Culture, (Princeton and Oxford: Oxford University Press, 2006), pp. 667-700

11. See Ken Saro-Wiwa, *On a Darkling Plain: An Account of the Nigerian Civil War*, (London, Lagos and Port Harcourt: Saros International Publishers, 1989) p. 51

12. It may be worth noting that at the time of independence, the smaller ethno-national groups collectively formed a significantly larger population than the largest majority group, the Hausa-Fulani.

13. See Ken Saro-Wiwa, *A Month and a Day*, p. 131

14. See, for example, Rob Nixon, "Ken Saro-Wiwa, Environmental Justice and Micro-Minority Rights" in Craig W. McLuckie and Aubrey McPhail Eds. *Ken Saro-Wiwa: Writer and Political Activist* (Boulder, Colorado and London, England: Lynne Rienner Publishers, 2000), pp. 109-125

Uyanga and many more—who inhabit the Niger Delta in the Eastern region and who came to be known during the colonial era as the "Rivers People". The Ogoni comprise just over one per cent of the total population of Nigeria which today stands at over one hundred and thirty million and speak four related, but mutually unintelligible, languages— Khana, Tai, Gokana and Eleme. They inhabit an oil- and gas-rich coastal area where the Ibo majority of the Eastern region has secured economic and political dominance.

Like many other Nigerian writers, Saro-Wiwa primarily used English rather than his first language, Khana, to make his views known both nationally and internationally—producing a total of twenty-seven books alongside numerous shorter pieces, such as poems, essays, speeches and newspaper articles, during his lifetime.[15] Before his involvement in environmental movements, he was best known in Nigeria as the creator of the hit comedy TV series, *Basi & Company*, which aired weekly on prime-time television between 1990 and 1995 to an estimated audience of 30 million viewers and out of which he also generated a popular series of children's books;[16] outside Nigeria, he had already garnered considerable attention within literary circles for his anti-war novel, *Sozaboy* (1987), and his short story collection, *A Forest of Flowers* (1987), both of which were first published in the Longman African Writers Series.[17]

Unlike other more internationally-renowned Nigerian writers, including, for instance, Chinua Achebe, Wole Soyinka or Ben Okri, however, Saro-Wiwa cannot easily be placed into the critical categories often used to describe West African writers and their preoccupations.[18] He came a little late to the literary scene to be counted among the "first generation" writers of the 1950s and '60s who were largely preoccupied with "writing back" to, and countering, colonial texts and traditions which had denigrated African cultures and

15. For a comprehensive bibliography of Saro-Wiwa's published work, see Craig W. McLuckie and James Gibbs, "Appendix 3: An Annotated Bibliography" in McLuckie and McPhail Eds. *Ken Saro-Wiwa: Writer and Political Activist*, pp. 245-284

16. See Nixon, p. 113

17. In 1987, *Sozaboy* had received an honourable mention in the Noma Award for Publishing in Africa and, in the same year, *A Forest of Flowers* was short-listed for the Commonwealth Writers Prize. See Laura Neame, "Appendix 1: Chronology of Ken Saro-Wiwa's Life" in McLuckie and McPhail Eds. *Ken Saro-Wiwa: Writer and Political Activist*, p. 234

18. 32 See Newell, pp.22-23

populations–often by dramatizing the devastating consequences for indige-nous populations of the colonial encounter (Achebe's *Things Fall Apart* is the classic example). But neither can he be accused of the alienation and detach-ment which is held to characterize the writing of the "second generation" as they faced the certain failures of African nationalist projects in the aftermath of decolonization in the 1970s. If he did not fall prey to the fearfulness and cynicism which the Kenyan writer Ngũgĩ Wa Thiong'o attributed to the latter cohort of writers in his classic polemic, *Decolonizing the Mind* (1986),[19] it was probably because he wrote primarily for local audiences rather than for a world market–as his son Ken Wiwa once put it in an interview, "'what [Saro-Wiwa] was interested in as a writer was reflecting the social conditions of Nigeria for Nigerians'".[20] Despite considerable successes in education, poli-tics and business which allowed him to transcend the relatively humble cir-cumstances of his birth and social background, he remained throughout his life informed about, and connected with, the material and social conditions and predicaments facing ordinary Nigerians. His commitment to making a public contribution in Nigeria even led him to establish his own publishing company there– Saros International Publishers–in 1973, when he became impatient with the slow pace of publishing in the West.[21] Rather than allow his success as a writer to depend upon the caprices of the Western market in educational books, self-publishing–sometimes at a financial loss–allowed Saro-Wiwa to pay more concerted attention to the difficulties endured by the Ogoni and other smaller ethnic nations whose position within the post-colonial state was persistently precarious.

In his memoir of the Nigerian Civil War, *On a Darkling Plain*, published by Saros in 1989, for example, Saro-Wiwa explains his opposition to secession and support for the federal government during the Civil War with reference to the relationships that obtained between the Ibo elite and the ordinary Ibo and other smaller ethnic nations, including the Ogoni, who inhabited the seceding Eastern Region, Biafra. Before the war began, he observes, "[a]

19. See Ngũgĩ Wa Thiong'o, *Decolonising the Mind: The Politics of Language in African Lit-erature* (Oxford: James Currey; Nairobi. East African Educational Publishers; Portsmouth, NH: Heineman, 1997), p. 22

20. See Laura Neame, "Saro-Wiwa the Publisher" in McLuckie and McPhail Eds. *Ken Saro-Wiwa: Writer and Political Activist*, p.160

21. Ibid. pp. 153-156

lmost 94% of the Region's crude oil production[...]came from the non-Ibo sections of the Region"[22] and, as a result, there had been considerable Ibo investment within the region outside Iboland and especially in the coastal cities of the Niger Delta. Ibo economic dominance in the oil-producing areas had also been consolidated, Saro-Wiwa maintains, by their control in the East, under the federal system, over offices within the civil service, government boards and corporations.[23] And while it was clear that the Ibo elite were disadvantaged by the Hausa-Fulani monopoly over government at the level of the state, he contends that they were not averse to using the numerical advantage of their ethnic group within the Eastern region to secure undemocratic gains there. In fact, he construes the greed of the Ibo elite as a key factor in the 1967 secession of the region from the federal state.

To say that the politics of the civil war were complicated would be an understatement. Numerous biographical, historical and fictional accounts have been produced but the causes and consequences of inter-communal conflict and violence—from the September massacre of 80,000-100,000 Ibos in 1966 to the end of the siege of Biafra in 1970, during which approximately 2.5 million Biafrans died—still merit further exploration.[24] While it has become clear, for instance, that Federal forces could not have laid siege to Biafra without the support of the British government and military—who, by the end of the war, had supplied 97% of federal arms— disagreement persists regarding the decision for the East to secede.[25] Some commentators continue to regard it as a visionary effort by the Ibo leadership to throw off the

22. See Ken Saro-Wiwa, On a Darkling Plain: An Account of the Nigerian Civil War, p. 53
23. Ibid. p. 54
24. In 1987, Craig McLuckie prepared a "checklist" of primary and secondary sources on the literature of the Nigerian Civil War which includes many works by Nigerian leaders (e.g. Nnamdi Azikiwe, Odumegwu Ojukwu) and major authors (eg. Chinua Achebe, Buchi Emecheta, J.P. Clark, Flora Nwapa, Christopher Okigbo, Wole Soyinka). See Craig W. McLuckie, "A Preliminary Checklist of Primary and Secondary Sources on Nigerian Civil War/Biafran Literature", Research in African Literatures, 18:4 (Winter, 1987): 510527. The war continues to preoccupy Nigerian writers in the contemporary moment. See, for instance, Chimamanda Ngozie Adichie's prize-winning novel, Half of a Yellow Sun (New York: Anchor Books, 2006) and Chinua Achebe's memoir, There Was a Country: A Personal History of Biafra (New York: Penguin Press, 2012)
25. See Herbert Ekwe-Ekwe, The Biafran War: Nigeria and the Aftermath (Lewiston, Queenston, Lampeter: The Edwin Mellin Press, 1991), p. 94

compromised state system bequeathed by colonial powers.[26] Others, including Saro-Wiwa, deplored the decision, characterizing General Odumegwu Ojukwu, who led the secession in 1967, as a short-sighted megalomaniac and gambler. "He was on a naked quest for power" Saro-Wiwa wrote in *On a Darkling Plain*, "and Ibo suffering at the time was his ladder to that power".[27].

While Saro-Wiwa did not hide his disdain for the Biafran leaders, however, his stance on the war in his memoirs, fiction and poetry is less stalwartly federalist than unfalteringly anti-war. As literary critic Charles Lock has convincingly argued, it is the seductive fervour of the Ibo leadership's self-righteous, war-mongering rhetoric more than the events of the war itself that forms the subject of his most acclaimed literary work, the experimental novel, *Sozaboy*.[28]

The interest of this novel for postcolonial literary critics first lay in its defiant use of a dialect, which Saro-Wiwa, in his prefatory "Author's Note" called "rotten English"[29], not simply in the dialogue between characters—or within what literary critics call the "inner frame" of the text—but throughout it, and even, as Lock points out, within the para-textual elements of the novel such as chapter headings.[30] However, while Saro-Wiwa was initially celebrated for granting an unofficial, spoken language the status and dignity of a literary one, upon closer examination of the text, the critical response became more cautious. There was no real attempt in the novel to mirror the language of everyday life nor to represent accurately the speech patterns of a particular population group. While Saro-Wiwa's "Author's Note" explains, for example, that the "rotten English" of his narrative comprised "a mixture of Nigerian pidgin English, broken English and occasional flashes of good, even idiomatic English", his protagonist Sozaboy [Soldier-boy] did not appear to engage in code-switching (where his choice of words, whether pidgin or standard English, could have been expected to vary depending on his audi-

26. 40 Ibid. pp. 120-121
27. See Ken Saro-Wiwa, *On a Darkling Plain*, p. 85
28. See Charles Lock, "Ken Saro-Wiwa, or 'The Pacification of the Primitive Tribes of the Lower Niger'" in McLuckie and McPhail Eds. *Ken Saro-Wiwa: Writer and Political Activist*, pp. 3-16
29. Ken Saro-Wiwa, *Sozaboy: A Novel in Rotten English* (Essex, England; New York; Mississauga, Ontario: Longman Publishing Group, Tenth Impression, 2006)
30. 44 Ibid. p. 9

ence or location).[31] Rather, the language of the narrative seemed as arbitrary and inconsistent as the process of decomposition announced in the word "rotten" itself.

Taking by way of example, however, an announcement by Shell– which appeared in many of the major English-language newspapers only eleven days after Saro-Wiwa's execution–in which the corporation presented itself as a body that is "Clear Thinking in Troubled Times",[32] Lock has convincingly argued that the author's rhetorical goal in *Sozaboy* was to expose the links between a modern "aesthetics of linguistic and stylistic purity" and "a technology of pollution".[33] For Lock, it is the "ethical confusion or pollution [of the political context] that accounts for and is reflected by"[34] the shifting and unpredictable language of the novel. What is under investigation in the novel, he argues, is less the realism or accuracy of any given account of the war and more the "order and rationality we [mistakenly] ascribe to *language* [Lock's italics]"[35]; what Saro-Wiwa seeks to convey are its fragility and limits.

For Lock, this point is delivered quite persuasively in the comical scene where Sozaboy first hears a military officer call for support for the war:

> The man with the fine shirt stood up. And begin to talk in English. Fine fine English. Big big words. Grammar. "Fantastic. Overwhelming. Generally. In particular and in general". Haba, god no vex. But he did not stop there. The big grammar continued. "Odious. Destruction. Fighting".
> I understand that one. "Henceforth. General mobilisation. All citizens. Able bodied. Join the military. His Excellency. Powers conferred on us. Volunteers. Conscription". Big big words. Long long grammar.[36]

Coming quite early in the narrative, Lock argues that this scene quickly alerts

31. See Doris Akekue, "Mind-Style in *Sozaboy*: A Functional Approach to Language" in Charles Nnolim Ed. *Critical Essays on Ken Saro-Wiwa's Sozaboy: A Novel in Rotten English* (London: Saros International, 1992), pp. 22–23. See also Michael North, "Ken Saro-Wiwa's *Sozaboy*: The Politics of 'Rotten English'", *Public Culture*. 13.1 (2001): 97-112
32. See Lock, p. 13
33. Ibid. p. 15
34. Ibid. p. 13
35. Ibid. p. 8.
36. See Ken Saro-Wiwa, *Sozaboy*, pp. 46-47

the novel's readers to the fact that Sozaboy "associates war with grammar"[37] so much so that, by the end, the "linguistic impurities" of the protagonist's informal, oral narrative themselves come to feel "cleansing" to the reader, from an ethical point of view.[38]

Of course, the broken-ness of this passage–its fragmentary sentences, absurd conjunctions and *non sequiturs* ("In particular and in general.")– also gives a literal quality to the broken relationship between the break-away military elite and the urban population of the Eastern region, subverting the claims of the officer in this scene to authority and legitimacy. Perhaps more than any other in the novel, the passage demonstrates the almost complete exclusion of the urban masses from the Biafran project of Afro-centric national self-development. Although he is located in some of the war's most dangerous theatres, much of the tragicomic hilarity of Sozaboy's story derives from his efforts to make sense of his wartime experiences with reference to the oral narratives of World War II that were conveyed to him by an African ex-serviceman in his hometown Dukana; he continuously confuses Biafra with Burma, Gowon with "Hitla" [Hitler].[39]

Since Saro-Wiwa's writings demonstrate such a cautious approach to leadership, it is perhaps ironic that, towards the end of his life, he found himself watching his own image assume a disembodied and iconic quality akin to the images of other political leaders who had been martyred for black nationalist causes, including Patrice Lumumba, or Malcolm X. Already in *A Month and a Day*, Saro-Wiwa expresses unease regarding the dehumanizing effects of his repeated appearance in the media–"I had been very much in the news lately and, as often happens to those who have the misfortune, was considered more as a news item than as a living being with flesh and blood".[40] And, in the following letters to Sr. Majella McCarron, we see him reject the static image of himself as hero or martyr in favour of the more active and open-ended figure of self as voice.

Doubting, for instance, that he will be released on time to attend the Right Livelihood Foundation ceremony planned in his honour in December 1994,

---

37. See Lock, p. 9
38. See Lock, p. 14.
39. General Yakubu Gowon, who was Head of State in Nigeria from 1966-1975 and led the federal forces during the Civil War.
40. See Ken Saro-Wiwa, *A Month and a Day*, p. 5

he concludes: "There or not, my words will ring through all the places";[41] just over three months later, some discussion of his trial on the BBC World Service and The Voice of America prompts him to remark that he has heard a version of himself reflected back on the airwaves: "The radio waves were full of the trial on the 6th February. The BBC World Service made it the first item in their news bulletins[...]The Voice of America also carried it fully. I heard myself described as 'renowned writer and environmentalist'".[42] Then again, when Shell, under pressure from MOSOP (acting in co-ordination with UNPO and other indigenous rights organizations), agreed in 1995 not to renew their operations in Ogoniland without the consent of the local population, he rejoices: "I hope also that other oil-bearing areas are listening. If they are, that should introduce a new situation into the Nigerian equation".[43]

The poignant sense Saro-Wiwa conveys in these letters—while confined, ill, and enduring harsh restrictions on his interactions with his family, friends and supporters—of having a voice that resounds globally echoes some of the grandeur of the rhetoric of much earlier twentieth-century Pan-African nationalists, including, for example, W.E.B. Du Bois or C.L.R. James, who often conceived themselves, in quasi-messianic terms, as the agents of a new and more just global social order.[44] But as Ato Quayson cautions in an important essay on the Ogoni crisis "African Postcolonial Relations Through a Prism of Tragedy", Saro-Wiwa's appeals to global frameworks should neither be regarded either as arrogant nor as perpetuating an out-of-date, Enlightenment era, universalist doctrine. In his essay, Quayson acknowledges that the Ogoni leadership had to operate within and contend with a social environment which had become "heavily over-determined by the ethos of Nigerian politics" and, in particular, by the "system of political patronage" established during the Babangida dictatorship 1985-1993.[45] He also recognizes that Saro-Wiwa frequently draws upon lofty and sublime

41. Letter dated 24th October 1994. MU/PP7/12
42. Letter dated 7th February 1995. MU/PP/7/22
43. Letter dated 2nd May 1995. MU/PP/7/25. Shell had suspended their operations in Ogoniland in 1993.
44. See, for example, W.E.B. Du Bois, *The Souls of Black Folk* [1903] (New York: Penguin, 1989); W.E.B. Du Bois. *Darkwater: Voices from within the Veil* [1919] (Minneola, New York: Dover, 1999); C. L. R. James. *Nkrumah and the Ghana Revolution* (London: Allison and Busby, 1977)
45. See Ato Quayson, *Calibrations: Reading for the Social*, p. 69

genres, such as tragedy and epic, to characterize the Ogoni situation. (One of the more striking examples is the Ogoni Bill of Rights drawn up by Saro-Wiwa in 1990 which, in echoing the American Bill of Rights and the French Rights of Man, aimed to endow the Ogoni cause with a world-historical, or global, rather than simply local or domestic character.[46]) But Quayson urges us not to view these rhetorical strategies as either self-aggrandizing or evidence of a desire on Saro-Wiwa's part "to assume the mantle of leadership [of MOSOP] as irrevocably his".[47] For him, Saro-Wiwa's disagreements with the more conservative members of MOSOP's leadership in the early '90s constituted a pragmatic response to what would later prove to have been a clear understanding of the likely conduct of the military regime rather than "a drive for personal authority".[48] The appeals Saro-Wiwa makes to global reference points and frameworks should not be interpreted as evidence of any faith in the idea of "natural leadership" or in universal principles but as efforts to re-configure and re-calibrate class arrangements within the historically short-lived state with attention to transnational and international processes.

In his letters here, Saro-Wiwa continues to use world-historical and sometimes planetary metaphors and language in order to give the Ogoni situation a recognizably international character. But it is worth noting that, when referring to universal or global frameworks and paradigms, he characteristically thinks "outside the box"; he repeatedly gestures, that is, towards what has been excluded or escaped. "Remember Hamlet?", he once gently reminds McCarron (when she appears to have become concerned about a possible threat he has made to practice juju against his captors), "There is more in Heaven and Earth than is compounded in your philosophy, Horatio."[49]

Indeed, even as he begins to recognize that his death is approaching, he resists the temptation to conceive himself as heroic and portrays himself, in a manner more in keeping with the language of *Sozaboy*, as self-divided or incomplete. There is, at times, a harrowing sense of self-dividedness here, that alternatively produces gallows humour—when, for example, he pro-

---

46. Ibid. p. 69
47. Ibid. p. 70
48. Ibid. p. 70
49. Letter dated 27th October 1994. MU/PP/7/13

fesses glee in anticipation of the "belly laughs"[50] that he expects state prosecutors will raise as they adopt absurd measures to bypass juridical process in order to convict him—and sadness—as when he first holds his youngest son Kwame and recognizes "I was not important to him".[51] And yet, somehow this sense of dividedness and incompleteness also provides Saro-Wiwa with a rhetorical basis for thinking about his relationships with others and how the Ogoni might rethink and reconfigure their relationships with other minority populations both within Africa and around the world.

Consider, for example, the autobiographical lyric poem "I Lie Alone At Night", probably addressed to Hauwa Maidugu dated 3rd June 1995, where, in the absence of the face of his beloved, the speaker is forced to consider the blank face of the moon. It begins with the stanza "I lie alone at night /And think all of one year's gone/Since I held you in my arms/In the bed we know so well." Six stanzas later, the speaker is still plaintively regretting his absence from his beloved's bed. Here though, the blank face of the moon strangely takes on the capacity to capture and reflect back something of the beauty of her far-away face: "I lie alone at night/And think of the stranger moon/The stars beyond my gaze/Your beauty like moons and stars."

50. Letter dated 21st March 1995. MU/PP/7/23
51. Letter dated 15th January 1995. MU/PP/7/21

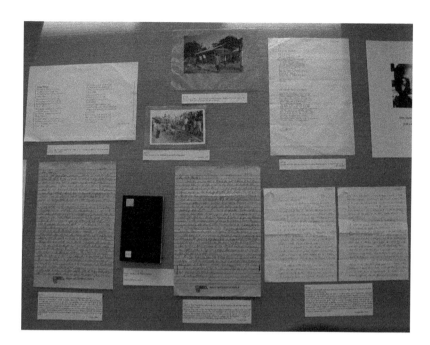

Letters and poems from Ken Saro-Wiwa. Maynooth University Ken Saro-Wiwa Archive.

In A *Month and a Day*, Saro-Wiwa had observed that:

> In virtually every nation state there are several 'Ogonis'–despairing and disappearing people suffering the yoke of political marginalisation, economic strangulation and environmental degradation, or a combination of these, unable to lift a finger to save themselves. What is their future?[52]

Here, as if in partial answer to that question, the idea that the blank, mirror-like face of a "stranger moon" viewed from a detention cell might reflect something of the beauty of a loved one's distant face is now extended to others, and in particular to the strangers, or anonymous individuals and populations, whose personal and intimate relationships have similarly been

---

52. Ken Saro-Wiwa, A *Month and a Day*. p. 123

broken by greed, totalitarianism and war, who anonymously endure "[b]roken hearts, breaking souls/[e]mpty dreams and lonely beds".

By conceiving the blankness of the moon as the reflection of the anonymity of others (who are nonetheless like himself), Saro-Wiwa, alone in his cell, then begins to conceive himself as one among many–perhaps hundreds of thousands, or even millions of people–and, in doing so, to discover the courage he needs to renew his contribution to the struggle for social change in Ogoniland and also in the more general pattern of power relations between other small ethnic nations in Africa and the global social order: "I lie alone at night/And dream a great new dawn/Without boots and knives[...]". The poem then ends with a new nocturnal vision, one of a night that, while still captured, is now "captured by peace". Saro-Wiwa's characteristic awareness of the fragmented, broken and incomplete character of imagery and words keeps language active and "alive" and prevents it from hardening into bureaucratic orthodoxies; it allows him to turn language inside out and to reverse it in order to chart new "cartographies of struggle"[53], new connections and relationships among human groups, new and more reciprocal relations of power among minority populations and the states they inhabit.

53. See Chandra Talpade Mohanty, "Cartographies of Struggle: Third World Women and the Politics of Feminism" in Chandra Talpade Mohanty, Ann Russo, Lourdes Torres Eds. *Third World Women and the Politics of Feminism* (Bloomington: Indiana University Press, 1991) pp. 1-46

Ken Saro-Wiwa in Political Context: Social Movements in the Niger Delta

LAURENCE COX

International solidarity between Africa and Ireland

THESE REMARKABLE LETTERS, from their writing to their publication, trace a history of international solidarity. Their origin lies in the relationship between Ken Saro-Wiwa, the wider Ogoni movement and Majella McCarron's solidarity work, well-documented in Helen Fallon's piece above. The letters themselves show Saro-Wiwa in action, mobilizing international support for himself and his co-defendants.

As he was only too aware, the situation of the "Ogoni Nine" symbolised the struggle between MOSOP, the Movement for the Survival of the Ogoni People, on the one side, and the Nigerian military government, together with the energy multinationals, on the other. To raise awareness and seek international solidarity for one was also to build support for the other. This strategy developed Saro-Wiwa's earlier internationalisation of the conflict through participation in the Unrepresented Nations and Peoples Organization and other fora.

When she returned to Ireland, Sr. Majella engaged closely with the Shell to Sea campaign of the equally remote community in Erris, County Mayo–the "Bogoni" as they sometimes called themselves–once again resisting Shell activities. As she made clear at the presentation of the letters, her choice of Maynooth University as a suitable repository for this historic collection is a result of the connections built up by Maynooth students working in solidarity with the campaign in Mayo, particularly John O'Shea, who suggested the donation; Jerrie Ann Sullivan, who recorded Irish police joking about the rape and deportation of protestors; and Terry Dunne, one of the original founders of the Rossport Solidarity Camp. Social movements can respect students and academics who take a stand on the basis of their own study and ethical choices; this engagement strengthens both civil society and intellectual work.

In turn, this publication would not have seen the light of day without the connection with Firoze Manji[1], the founder and former editor-in-chief of Pambazuka News, the pan African website and e-newsletter, who saw the significance of the letters at once and whose commitment to the project of publishing these letters has been exemplary. With this publication, the connection between African and Irish social movements comes full circle. But why should these movements exist in the first place?

The curse of oil: between Norway and Nigeria

The Norwegian specialist on oil and gas, Helge Ryggvik, speaks of the "curse of oil". Finds of gas and oil only rarely benefit the local population. More commonly they lead to greater social and economic inequality, the corruption of the state apparatus and increased violence, as the rich and powerful use their existing advantages to monopolise the benefits of the new wealth.[2]

Norway, with all its complexities, is a rare exception to this rule. This is not because of any natural facts but because popular movements were determined that oil and gas wealth would be used for the benefit of society as a whole. In particular, civil servants who had been part of the resistance to fascism during the Second World War, an assertive trade union movement and popular commitment to social equality were crucial in pushing through the "Norwegian model" against the resistance of conservative elites and threats from the energy companies.

The essence of this was a productive model which gave multinationals limited access in the early phases in return for skill and technology transfer, which enabled Norway to build up its own industry, along with ownership

1. Firoze Manji, founder and former editor-in-chief of Pambazuka News, was head of CODESRIA's Documentation and Information Centre <http://www.codesria. org>. He is the publisher of Daraja Press (https://darajapress.com).
2. Helge Ryggvik, *Norway's Oil Experience: A Toolbox for Managing Resources?* Oslo: Centre for Technology, Innovation and Culture, University of Oslo, 2010. English translation available at <http://tinyurl.com/norwegianoil>

and tax regimes which led to the Norwegian state's petroleum fund becoming one of the world's largest pension funds and financial investors.

Ireland and Norway both became independent from neighbouring powers in the early twentieth century; but Irish politics took different routes to the Norwegian, leading to the country becoming one of the most unequal in western Europe, with notoriously high levels of state corruption and clientelist power relations. It is this, rather than (as government ministers sometimes claim) some unique physical disadvantage vis-a-vis Norway, which has shaped Ireland's response to oil and gas: a long-standing political commitment to enticing multinationals at any cost, coupled with the determination of local elites to secure small-scale advantages as middlemen in what has recently been called a "meet-and-greet" capitalism,[3] and the defeat and subordination of popular social movements.[4]

As the Navy is used to force through the Corrib pipeline in Mayo, further exploration rights are given away off the west coast, fracking is threatened across the west Midlands, and oilfields are announced off Dublin and Cork—all against a backdrop of IMF- and EU-mandated austerity and media cheers—the Irish future seems likely to be closer to Nigeria than Norway.

The Niger Delta and the Movement for the Survival of the Ogoni People

The people of the Niger Delta have suffered massively from their location on top of Nigeria's oilfields. Their history, as Ken Saro-Wiwa wrote, has been one of a massively accelerated colonisation—the first British policemen arrived to impose imperial control as late as 1901 and conquest was only complete in 1914—tied to the then-crucial palm oil trade.[5]. Following the start

3. Conor McCabe, *Sins of the Fathers: Tracing the Decisions that Shaped the Irish Economy* (Dublin: The History Press, 2011)
4. Laurence Cox, "Gramsci in Mayo: a Marxist Perspective on Social Movements in Ireland", Paper to New Agendas in Social Movement Studies Conference, Maynooth (2011) <http://eprints.nuim.ie/2889/>
5. Sanya Osha, "Birth of the Ogoni Protest Movement", *Journal of Asian and African Studies.* 41.1/2 (2006): 13-38

of oil exploitation in the 1950s, multinational companies led a partial industrialisation, intertwined with internal colonisation by the newly independent Nigerian state.

MOSOP *protest in Ogoniland. Maynooth University Ken Saro-Wiwa Archive.*

Indigenous groups, in particular, suffer massively from their location in what are often strategic areas for extractive industries.[6] Ogoni are only one of several affected populations in this large and ethnically complex area. Such populations have truly been cursed by oil and gas: the vast bulk of Nigeria's GDP and virtually all of its export revenue is generated here, where less than a quarter of the population live, and they are thus caught between the multinational energy companies—some of the world's most powerful actors—and the national state.

The inequalities of high-tech industrial development side-by-side with rural poverty, the environmental devastation of blowouts, oil spills and the world's largest flaring of natural gas, political marginalisation and police and

6. Declaration of the International Conference on Extractive Industries and Indigenous Peoples, Manila, 2009

military brutality have been the norm rather than the exception for indigenous groups in the Delta. From the Biafran war through to the current militarisation of the Delta, struggles for the control of the vast wealth represented by oil have taken place at the expense of local populations.[7]

As the newly independent African states' pursuit of national economic development produced increasingly uneven distributions of wealth, repression of popular movements and ultimately acceptance of the "adjustment" packages of International Monetary Fund / World Bank austerity, poor rural groups and ethnic minorities suffered disproportionately from the deregulated economy. In the 1980s in particular, environmental movements revolving around the key resources of land, water, forests and so on challenged the state and multinational capital and connected this to the wider struggle for a people-centred democracy.[8]

MOSOP—the Movement for the Survival of the Ogoni People—was unusual in the history of resistance to this process in that it was determinedly democratic and non-violent; a truly popular mass movement aiming to secure popular control of this wealth, which did not simply reproduce the state's own practices of violence.[9] As a federation of Ogoni organizations, including

---

7. For an overview, see John Agbonifo, "Territorialising Niger Delta Conflicts: Place and Contentious Mobilisation", *Interface: a Journal for and about Social Movements* 3.1: 240-265 <http://www.interfacejournal.net/wordpress/wp-content/uploads/2011/05/Interface-3-1-Agbonifo.pdf>

8. Cyril Obi, "Environmental Movements in Sub-Saharan Africa: a Political Ecology of Power and Conflict", UN Research Institute for Social Development, Civil Society and Social Movements programme paper 15 (2005), online via <www.unrisd.org>. For wider overviews of social movements in Africa, see the *Journal of Asian and African Studies* special issue, "Political Subjectivities in Africa" 47.5; Peter Dwyer and Leo Zeilig, *African Struggles Today: Social Movements Since Independence* (Chicago: Haymarket, 2012); Miles Larmer, "Social Movement Struggles in Africa", *Review of African Political Economy* 37.125: 251-262; Nikolai Brandeis and Bettina Engels, "Social Movements in Africa", *Stichproben: Wiener Zeitschrift für kritische Afrikastudien* 20: 1-15, <http://www.univie. ac.at/ecco/stichproben/20_Einleitung.pdf>; and Michael Neocosmos, "Civil Society, Citizenship and the Politics of the (Im)Possible: Rethinking Militancy in Africa Today", *Interface: a Journal for and about Social Movements*, 2.1: 263-334, <http://interfacejournal. nuim.ie/wordpress/wp-content/uploads/2010/11/Interface-1-2-pp263-334-Neocosmos. pdf>

9. Michael Neocosmos, "Transition, Human Rights and Violence: Rethinking a Liberal Political Relationship in the African Neo-colony", *Interface: a Journal for and about*

women's and youth groups, churches and traditional leaders, students and teachers, it had particular legitimacy. In early 1993, something like 60% of the entire Ogoni population joined a coordinated day of protest–a virtually unheard-of level of participation.

MOSOP cap and flag. Maynooth University Ken Saro-Wiwa Archive.

Social Movements, 3.2: 359-399, <http://www.interfacejournal.net/wordpress/wp-content/ uploads/2011/12/Interface-3-2-Neocosmos.pdf>

The Ogoni Nine

The military response by the then dictatorship was one of characteristic brutality: demonstrators were shot, roads were sealed and villages were destroyed.[10] The trial of Ken Saro-Wiwa and the other members of the "Ogoni Nine" was part of this process. In 1994, four Ogoni chiefs were killed under circumstances which remain unclear.[11] Fifteen Ogoni men were charged with the murders, including Saro-Wiwa, who had been refused entry to Ogoniland on the day of the murders.

As the letters show, it was clear to Saro-Wiwa from early on that the court was rigged[12] and the intention was to execute MOSOP's leaders in the belief this would end the movement. In fact, nine of the defendants, including Saro-Wiwa, were hanged, but the movement did not come to an end.

Instead, it was the dictatorship which fell in 1999.[13] State violence, it turns out, does not always win, and international outcry can make a dif-ference.[14] Nor did Shell win; it was declared *persona non grata* by MOSOP in 1993, sub-

---

10. Elowyn Corby, "Ogoni People Struggle with Shell Oil" (2011), Global Nonviolent Action database, <http://nvdatabase.swarthmore.edu/content/ogoni-people-struggle-shell-oil-nigeria-1990-1995>

11. *The Independent* (London), "Ken Saro-Wiwa was Framed, Secret Evidence Shows". 5th December 2010

12. See the highly critical report on the trial for the Article 19 human rights NGO by Michael Birnbaum QC, "Nigeria: Human Rights Denied. Report of the Trial of Ken Saro-Wiwa and Others". Article 19 (1995) <http://www.article19.org/data/files/pdfs/publications/nigeria-fundamental-rights-denied.pdf>. This notes among other things that the two principal witnesses subsequently swore affidavits claiming that they were bribed to give false evidence.

13. On the wider relationship between ethnic movements and democratisation, see Kehinde Olayode, "Ethno-nationalist Movements and Political Mobilisation in Africa: the Nigeria Experience (1990-2003)", *Stichproben: Wiener Zeitschrift für kritische Afrikastudien*, 20: 69-93, <http://www.univie.ac.at/ecco/stichproben/20_Olay-ode.pdf>

14. Marie-Emmanuelle Pomerolle, "The Extraversion of Protest: Conditions, History and Use of the 'International' in Africa", *Review of African Political Economy*, 37. 125: 263-279, rightly notes that internationalisation has costs as well as benefits; but few movements facing repressive regimes anywhere in the world have failed to appeal to movements and opinion abroad.

sequently left Ogoniland and eventually saw its operatorship handed over to the Nigerian National Petroleum Corporation in 2011. A somewhat greater share of oil resources now go to Ogoniland, and the Ogoni movement continues.

Saro-Wiwa's family, meanwhile, together with that of the other executed men, spent years pursuing Shell through the courts until they settled for over $15 million dollars. Court documents which have since become available make clear the extent to which Shell funded and instigated military operations against Ogoni activists in the Delta.[15] More recently, Wikileaks documents show Shell leadership boasting of their staff placed within Nigerian government departments and of exchanging intelligence with the US about militant activity in the Delta.[16]

Ken Saro-Wiwa and the letters

Ken Saro-Wiwa's prison letters, like his son's critical account[17], show him to be a genuinely remarkable figure and a social movement leader of great stature, who could draw on a breadth of political experience and historical vision. It is not too much to say that they bear comparison with the prison letters of the Italian anti-fascist leader Antonio Gramsci, which for decades have been staple reading for Italian schoolchildren.

At Gramsci's trial, the prosecutor famously stated that "we" (fascists) must keep this brain from working for twenty years. Saro-Wiwa's letters show his brain working at full speed, directing the campaign to halt the executions, developing strategy for MOSOP, and discussing the weakness of the Nigerian dictatorship. There is not only a clear analysis of the situation but also a hugely energetic activity directed not to saving his own life but to making

15. *Guardian* (London), "Shell Oil Paid Nigerian Military to Put Down Protests, Court Documents Show", 3rd October 2011
16. *Guardian* (London), "Wikileaks Cables: Shell's Grip on Nigerian State Revealed", 8th December 2010
17. Ken Wiwa, *In the Shadow of a Saint: a Son's Journey to Understand his Father's Legacy* (London: Black Swan, 2001)

sure that a death—which he in part expected—would serve the Ogoni movement and the end of dictatorship—as indeed it did.

Like Gramsci's letters, Ken Saro-Wiwa's prison letters deserve a wide readership. They are immensely readable, hugely enlightening and inspiring for all those who struggle for social and political justice. They show that even the world's largest energy companies can be defeated; that even the most brutal and corrupt regimes do not necessarily win; and that even the smallest and most marginal populations, when they have their backs to the wall, a commitment to organising themselves and allies from other social movements, can find reasons to hope and act.

Your Ogoni, my Fermanagh

One of Saro-Wiwa's poems, for Majella McCarron, sums this analysis—and his internationalism—up perfectly: "... *reach out to the grassroots / Of your Ogoni, my Fermanagh*". Ogoni's situation is ours, and our situation is theirs; we can learn from each other.

The struggle for the democratic control of oil and gas resources, to ensure that when extracted their use is to benefit society as a whole and that only when appropriate ecological considerations prevail, is as global as the energy multinationals. Oil and gas go hand in hand with corrupt states, thuggish police forces, judges who do what they are told and a bought media.

They do not only bring out the worst in people, though; communities standing on their own land and fighting for their future existence often show extraordinary courage in the face of this intimidation, great generosity towards similar struggles elsewhere, a poetic vision of how we *should* live and a clear perspective of the realities not only of power but of the longer timescales in which we have to think if we want to have grandchildren and not simply bank accounts.

Ken Saro-Wiwa, in the letters collected here, exemplifies what human beings can be, in a standing reproach to the short-sighted, the vicious and the corrupt and a lasting inspiration to the wise, the decent and the brave.

The Maynooth University Ken Saro-Wiwa Audio Archive

DR ANNE O'BRIEN AND HELEN FALLON

KAIROS COMMUNICATIONS, under the direction of Dr Anne O'Brien, filmed the handover of the Ken Saro-Wiwa collection to Maynooth University in November 2011. From this and subsequently conversations the concept of an audio archive took root.

This is the story of the Ken Saro-Wiwa Audio Archive, which was created by Dr Anne O'Brien and Helen Fallon. The archive is freely available online. It has been accessed over 2,000 times since its creation and continues to be a living archive of people and events related to Ken Saro-Wiwa and his ideals.

Recordings

As of August 2018 there are 15 recordings in the audio archive and these can be accessed at:

https://www.maynoothuniversity.ie/library/collections/ken-saro-wiwa-audio-archive

Recordings 1-8 – The story of Sister Majella McCarron, from her childhood in rural Fermanagh to her work on social justice issues today.

Recording 9 – Dr Íde Corley discusses Ken Saro-wiwa's position in post-colonial African literature and his role as a popular novelist and the creator of an award winning television sitcom.

Recording 10 – Helen Fallon discusses the Maynooth University acquisition of the Ken Saro-Wiwa collection.

Recording 11 – Dr Laurence Cox discusses Ken Saro-Wiwa's legacy in terms of conflict over natural resources and the importance of the archive for both researchers and activists.

Recordings 12-14 – Dr Owens Wiwa talks about growing up in Ogoni, the environmental destruction of his homeland and the deaths of his brother and his colleagues. In recording 14, he reads two of his brother's poems.

Recording 15 – An interview with Noo Saro-Wiwa.

While the archive of letters and images that McCarron donated to the library is immensely valuable, and the book *Silence would be Treason* offers an astute analysis of Saro-Wiwa's work from a variety of perspectives, neither the archive nor the book entirely tell his story. The audio archive was designed to set out a more detailed and direct account of that story. Usually the hardest part of creating an audio archive is deciding on where the story is to be found. This was not the case with the Ken Saro-Wiwa audio archive, as the people who had the story were quite obvious, available, and willing to participate. In this case the story was most closely observed, from an Irish perspective by McCarron, and from a personal perspective by Saro-Wiwa's brother, Owens Wiwa. The audio archive also contains an interview with Dr Íde Corley on post-colonial African literature; Helen Fallon talking about the importance of the letters and other materials in the collection, Dr Laurence Cox addressing the environmental issues and Noo Saro-Wiwa speaking about her father and life in Nigeria and London. These recordings complement their input to the written volume of letters and poems.

Arguably, it is the unique contribution of the audio archive that two of the people who worked most closely with Saro-Wiwa – Sister Majella and Owens Wiwa – were given the opportunity to speak for themselves, directly and in as unmediated a manner as was possible. Both of these people had so often had their version of Saro-Wiwa's story told on their behalf, that it was important for them to have the opportunity to speak for themselves in an unconstrained manner. The absence of intervention in the telling of their versions of Saro-Wiwa's story became a guiding principle for the production of the audio recordings. This allowed the contributors to maximise their imprint on the story, centrally guiding the subsequent editing process.

The key freedom involved in creating the audio archive, as opposed to working to the format, genre and scheduling constraints of broadcast documentary or feature programming, was that it offered as much 'space' as is required to tell the story in the fullest detail possible. While broadcast programmes limit the on-air time allocated to a programme and so limit the amount of material recorded at source, this was not the case with an open-ended time allocation. Seven hours of audio were recorded with McCarron.

Majella McCarron, courtesy of Sister Majella McCarron.
Maynooth University Ken Saro-Wiwa Archive.

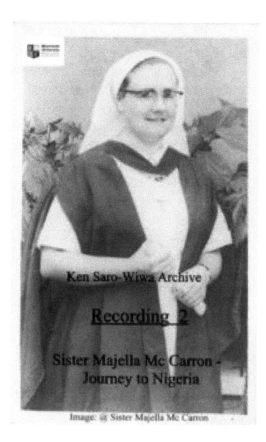

Sister Majella McCarron, courtesy of Sister Majella McCarron. Maynooth University Ken Saro-Wiwa Archive.

This was free-ranging in the topics covered, from her Irish childhood, her education, her missionary work in Nigeria, the events that brought her to Saro-Wiwa, and all that passed subsequently. Hearing her story told in her own voice offers an insight into her personality and character that was not always as immediately conveyed in the written word. Moreover, hearing her voice first hand, with the intimacy this creates in recounting events in Nigeria leading up to Saro-Wiwa's death, provokes a compelling intellectual and emotional awakening to the horror of the environmental abuse and destruction of Ogoni that she experienced first-hand.

Similarly, with Dr Owens Wiwa there was as much time available as needed to account for events in Ogoni and to tell the story of what had happened to his brother. Again, editorial intervention was minimised by recording the interviews as if they were being broadcast live. The interview with Owens Wiwa was recorded in a studio and there were no 'retakes' on any of the questions posed or answers offered. The interview as it exists in the archive is identical to that recorded in studio. This gave control of the final product, the archived version of the story to Owens Wiwa, rather than to an outside editor at a later point in time. In this way, by avoiding the possibility of editing, the audio archives can offer a safe 'home' to a story, a place where despite exclusion or misrepresentation in wider media, a story can be told and held with a minimum of editorial interference.

The interview with Noo Saro-Wiwa was recorded in a studio on her visit to Maynooth University on the 20th anniversary of her father's execution.

Noo Saro-Wiwa and Sister Majella Mc Carron. Maynooth University Ken Saro-Wiwa Archive.

She movingly told the story of how she found out about the death of her father:

> I was a second year university student at King's College London and he was sentenced to death – my father and his colleagues – on October 31st. So that came as a real shock, but then the international community really rallied round, so come November 10th it was just another day within that particular period. I must have attended classes, and then I went and did some shopping and I came back to the house I was living in in North London and my housemate had left a message, just a handwritten note on the table saying 'call your mother' and so I called my mother and she was the one who told me. I just put down the phone, which was the same reaction I had when I was told that my little brother died two years previously, I just put down the phone. And then went home immediately to my mother's house and spent the evening with the family – my cousins and my aunt and uncle came over."

As to the mechanics of producing the audio archive, it is in essence a series of recorded interviews. Many were recorded on location. The two authors visited Rossport in Mayo with Sister Majella for the reinstatement of commemorative crosses for the Ogoni Nine at the Bellinanboy Shell terminal in November 2012.

Sister Majella McCann – The Shell to Sea Campaign. Maynooth University Ken Saro-Wiwa Archive.

Other interviews were recorded from the comfort of McCarron's home, or in a car when nowhere more conveniently quiet presented itself. The aim in deciding the location for interview recordings was to make sure the sound was as 'clean as possible' to minimise distractions from the trajectory of the story itself. The interviews with Dr Owens Wiwa were recorded 'as live' in studio, when he came to Maynooth University to launch the book of his brother's letters.

Dr Owens Wiwa views his brother's letters at Maynooth University Library. Maynooth University Ken Saro-Wiwa Archive.

The interviews are, on first encounter, deceptively simple. They seem to meander through the story as if it is being told for the first time. This impression belies the volume of research that underpinned the detailed understanding of the story that the producers had acquired, the work involved in formulating and reformulating questions, and the care and time taken in conducting the interview on the day.

The recordings with Owens Wiwa attempted to follow best practice for interviews. There was no discussion of his brother or the 'story' prior to the recording, so that he did not feel he had already told the story, and might perhaps gloss over details in the interview. The session started with easy warm-up questions, such as childhood memories of his brother. This allowed the interviewee to settle into the interview and to get a feel for the right pace and tone of questions. From there the questions emerged from the chronology of events in Ogoni, leading up to Saro-Wiwa's final detention. All the time, the interviewer needed to keep Owens Wiwa on the trail of the story while watching and reading how far he was able and willing to relive the detail of the destruction of Ogoni and in particular his brother's death.

As the events recounted became more brutal and savage, the questions got shorter and simpler, relating to specific places and violent incidents – *what happened at Biara? What happened in Kaa? Tell me about Oloko?* These are simple questions that drove the narrative and were, for Owens Wiwa, the key turning points of the story; they shaped the topics that were discussed in the interview and gave the archive a route through a detailed and complex story that was coherent and manageable for non-expert listeners. Without thorough research and close listening, interviews don't always yield the kinds of key personal insights that Owens Wiwa so generously offered. Despite close research, it is not unusual for an interviewer to be surprised by answers, for instance when questions don't reveal the kind of insider tragedy or intimacy that was expected. When asked about the last time he had seen his brother Owens Wiwa gave quite an everyday answer, as if there was nothing of sentimental significance for him in the last meeting with his brother. In that case the unexpected answer is a useful reminder not to impose too much meaning or significance on events before the interview, but to remain alert to the dialogic exchange and possibilities that arise in the live telling of a story.

The archive required work in the pre-production phase, to gather as much information as possible on the story of Owens Wiwa and McCarron's relationships with Saro-Wiwa, and to formulate questions that allowed for broad and wide-ranging responses, while still carrying the story forward in a manner accessible to non-experts. In production the recordings involved long sessions, so that the final archive would offer unlimited space to the contributor's testimony. In post-production the editing was minimised, so that even apparently unrelated material was retained and valued as offering insights to McCarron's character and presence. Hopefully, the archive can become a place in which a story can be 'laid to rest,' where it can reside indefinitely. McCarron commented that the archive could act as an overall record of her life and what had happened to Ogoni and to Saro-Wiwa. To that end, it is important that the audio archive is publicly available and accessible.

The Ken Saro-Wiwa audio archive, which was launched by Dr Owens Wiwa, alongside *Silence Would be Treason*, can be accessed online.

Between the launch on 7th November 2013 and 31 August 2017, the audio archive has been played over 2,000 times and has been widely promoted. At lectures and presentations on the audio archive, the audience never fail to be moved by the sound of Owens Wiwa's voice, reading Ken's poems *Ogoni! Ogoni!* or *For Sister Majella McCarron*. While academics and activists have spoken about Ken Saro-Wiwa in their research, through their publications and in their teaching, it is in the timber and accent of Owens Wiwa's voice that Ken Saro-Wiwa really comes to life.

The Ken Saro-Wiwa Audio Archive provides an opportunity for students and the public to engage meaningfully with a complex and controversial topic that may seem very removed. There's an old saying that 'the pictures are better on radio.' In the case of the Ken Saro Wiwa audio archive, that saying holds true. People listening to the recordings can construct mental images of the lives of the key protagonists in Ken Saro-Wiwa's story and understand better the roles that Sister Majella McCarron and his brother Owens played in his life. There are no actual pictures to distract the imagination and so the listener can create their own landscape in an imagined Ogoni. But listeners don't just think in terms of pictures; audio allows the user to access the part of the mind that generates dreams, to conjure more than a three-dimensional picture of Ogoni. Audio allows the listener to smell, feel and taste the world it creates. Listeners to the Ken Saro Wiwa audio archive can smell the gas flares, taste the polluted water and the touch the oil-encrusted, infertile spoiled land. In so doing they can clearly understand why Ken Saro-Wiwa created the Movement for the Survival of the Ogoni People. The audio archive brings home the fact that it was, and is, the survival of the people that was at stake.

Beyond Ogoni, the audio archive helps the listener to understand other issues too. In the immediacy of the first-person accounts and the intimacy of listening to another human voice, the passing of time collapses, the relevance and reality of events from decades ago become immediate. The audio archive, in this way, offers an insight into religious formation and convent life in 1950s Ireland. It gives an insider account of the thoughts and feelings of young missionaries travelling to Nigeria in the 1960s. Through Majella McCarron's and Owens Wiwa's recounting of their time with Ken Saro-Wiwa, particularly in the 1990s until his untimely death, the users are allowed to 'see' Ken Saro-Wiwa, not just as an activist who paid the ultimate price for his beliefs, but also as a friend and brother. Listening to the recordings, the users come to an understanding of the invisible ties that bind Sister Majella's work in Nigeria with her activism today, against Shell in Erris, Co Mayo, in the West of Ireland. Somehow sound allows the listener to see very clearly the strong threads that connect the orator to the listener; the threads that also connect Ogoni to Ireland.

While recording that track of poetry for the archive, Owens Wiwa was asked if he would read from his brother's letters. He refused, explaining that he felt to do so would be inappropriate as it would be to speak in his brother's voice. On reflection, the most poignant aspect of the audio archive is the silence that lies at the heart of it, the listener doesn't get to hear Ken's voice. Throughout his life Ken fought for the rights of the Ogoni, and proclaimed that silence, in the face of their plight, would be treason. The stark fact remains, however, that Ken Saro Wiwa was silenced. That silence is in evidence amongst the voices of his 'Sister Majella' and his brother, Owens. It is a silence that offers testimony to the injustice and plight still suffered by Ogoni. Creating the Ken Saro-Wiwa Audio Archive is, the authors hope, a way to ensure the Ogoni story is not forgotten.

The Maynooth University Ken Saro-Wiwa Bursary: My Research on the origins of the oil industry

GRAHAM KAY

IN AUTUMN 2015 I WAS THE RECIPIENT of the inaugural Maynooth University Ken Saro-Wiwa bursary. My interest in Saro-Wiwa's life and works emanated from research I am conducting in the History Department at Maynooth University. My PhD thesis entitled 'The First Oil War: Britain, Germany, and the Race for Oil, 1896-1921', seeks to trace the origins and evolution of policy which led to a technological transformation and transition from coal to oil.

From early government measures to recognition as a vital strategic resource, and from driver in societal change to economic necessity, the development of oil is a fascinating, complex and curiously unexplored history. In a world that has now become obsessed with its acquisition, a complete understanding of the role of oil in strategic and political decision-making remains elusive. Today, oil has become fundamental to global infrastructure and those who control its supply possess the ability to arbitrarily influence world markets, but few have attempted to trace its origins as a factor in the development of policy and international affairs. My transnational and comparative project proposes the theory that the twentieth century saw the earlier focus on manpower replaced by a new paradigm: the belief that the acquisition of superior forms of energy revolutionised industry, and therefore military strength, allowing for a concept which would become known as 'Total War'.

Over the past 100 years, oil has weaved itself into the very fabric that maintains our society as we know it. From fertilizers to plastics, and credit cards to cigarette filters, oil has placed the world's economy in the difficult position of almost complete reliance, and with that dependence the importance of control emerges. My research attempts to confront the relationship between government and corporation with respect to a crucial resource underpinning our world today. By looking at the genesis of this relationship it will help us to understand how a trickle of influence grew to be an overwhelming force in the formation of policy. This force would eventually lead to the creation of major corporations with unparalleled financial backing to acquire the necessary agreements for further exploitation. Saro-Wiwa's life's work highlights this threat to, not only the environment, but also to how money, power, and natural resources influence a people or way of life. In my view, it is essential to examine where this record began so we can further recognise the significance of Saro-Wiwa's achievements. Indeed, his life and death underscores the need for us, the global community, to continue to understand and probe the societal impact of the oil industry.

The founding of the Anglo-Persian Oil Company (later British Petroleum or BP as they are more commonly known) and its acquisition by the British government in 1914 set in motion a scramble to ensure a strong and secure supply of oil for the Great Britain. In turn, this spurred other emerging oil companies such as Royal Dutch Shell, Standard Oil and the Burmah Oil Company to seek out and acquire vast areas to explore and exploit or fear falling behind the competition. Where a national interest existed, such as in the cases of Great Britain and Germany, pressure was placed on territories and nation states where oil could be accessed to give concessions and guarantees for specific companies, like those of Royal Dutch Shell and BP. As oil became linked to economic prosperity and military strength through advanced technology, it took on a very prominent political dimension. Demand for this emerging and superior source of energy grew rapidly with powerful states seeking to ensure they remained strong in a transforming world.

With Great Britain installed as the dominant naval power, though waning, as some authors would argue, in Europe in 1900, the discovery of oil as a source of energy levelled the playing field for those who could harness and develop technology which used it, and, in turn, rival powers sought to challenge the status quo. Germany, seeking to improve its political position on the continent, was gifted an opportunity to confront British ascendency. As industrial rivals, Great Britain and Germany initiated a game where both were forced to assess the significance and utility of an emerging resource in an age of rapidly developing technology. A comparison of how both states came to understand and utilise this resource is essential to producing a definitive account of their relationship before the war, their wartime attempts to expand into significant areas of the globe, and of the post-war implications for Europe and the Middle East.

Domestically, both states entered into programmes for research into the potential of oil-based technologies. With Great Britain and Germany spending ever increasing amounts of time and vast sums of money to understand how best to exploit this emerging energy resource, policies introduced by both governments suggested a growing relationship with a number of oil companies. As mechanisms for helping to steer foreign policy, oil companies operating in the Middle East and in colonial territories helped to solidify a national presence there and ensure an important element of security in an area of interest. The origins of this practice, and the purpose of this project, can be clearly observed, and remain evident today.

Ken Saro-Wiwa's political activism centred on environmental destruction and the effects of a large multinational on government policy – effects which are still resonating in present day Nigeria. My work focuses on the origins of a framework, which undoubtedly, led to a competitive structure of exploitation by developed nations, desperate for access to oil, on developing states. As such, this analysis and evaluation will be on the how and why government policy at this crucial juncture in time became so intrinsically linked to the issue of oil. By examining the beginning of this relationship, which effectively started in Great Britain and Germany due to, primarily, military rivalry and conflict, my research establishes a definitive approach for looking at the interaction between governments and corporations in both large and small oil producing countries.

What follows here is a detailed account of how the Maynooth University Ken Saro-Wiwa bursary enabled me to go in search of answers to difficult questions I posed in my research. Most of the answers I sought could only be found at the *Bundesarchiv* in Berlin, Germany. The records held by the State Archive was integral to my investigation, and as such, crucial to producing a comprehensive analysis of the wider implications of the early history of the oil industry in Germany. Before I left for Germany, however, I spent a few days in the National Archives in Kew in London. After the Second World War many German documents and records were taken by the Western Allies back to the United Kingdom and the United States. Copies of these documents, relating to many aspects of the German government from the early 20th century until the end of the Second World War, are now kept in the National Archives. The specific documents made available to me involved letters and memoranda from the German Foreign Minister sharing his concerns over what was to happen to the German oil industry after the war. More precisely, the primary concern was whether or not Germany, in the aftermath of 1914-1918, would be able to access a quantity of oil commensurate to the needs for rebuilding a devastated economy. The war had shown that while coal had, and continued to be, a suitable and abundant resource for industry, oil was found to be superior and without any doubt, would open up the frontier for more efficient and powerful technology. The concerns shared by the German Foreign Minister indicated an emerging policy which had transformed from an important issue before the war, to something fundamental to the reconstitution of the post-war German economy.

After London, I made my way to the *Bundesarchiv* in Lichterfelde – a suburban district not far from the city centre – in Berlin. The archive is in an old army barracks occupied by U.S. forces in Berlin up until the 1950s. By comparison to the National Archive in London, it is quite dated and unfortunately appears to receive very little funding. It was here, however, that most of the records and documents on the German Chancellery, the Foreign Ministry, the War Oil Department, and the Reich Economic Council were to be found.

Although I went to Germany with specific questions requiring clear answers, it didn't take long before I discovered additional lines of enquiry. Initially, I conducted research on the subject of Romania's participation in the First World War. At the time, Romania was one of the largest oil producers in the world and given her proximity to Germany and Austria-Hungary, coupled with her neutrality status up until 1916, I wanted to know how the Entente allies and the Central Powers engaged with the government in Bucharest. Romania's opportunistic entry into the war in the latter half of 1916, spurred by signs of a strong Russian offensive and Anglo-French efforts during the Battle of the Somme, was a grave misstep in calculation, short-sighted, and ultimately, a complete disaster. The consequences of this blunder led to the 'greatest single act of economic warfare in the war.'[1] In a bid to prevent Germany occupying and utilising the oil industry of the world's fifth largest oil producer, the Romanian government agreed to allow a British taskforce to destroy and make obsolete as much of their industry as possible. What followed was a whirlwind of sheer destruction which disabled the vast majority of Romania's oil producing refineries, drill sites, and transport services. General Ludendorff of the German army later chronicled his thoughts on John Norton-Griffiths' exploits: '...materially reduce[d] the oil supplies of our army and the home country' and that: '...we must attribute our shortages

---

1. Jay Winter (ed.) *The Cambridge History of the First World War*, Vol. II, (Cambridge, 2014), p. 477.

in part to him.[2] What the records in Berlin revealed was that the German forces, which later occupied Romania, invested a large amount of time and effort attempting to rebuild the crippled oil industry. In short, I discovered detailed files listing the exact weekly quantities being exported back to Germany from Romania, with a clear increase in output from January 1917 until the end of the war. My calculations show that in a little over two years, German forces were only able to restore output to forty percent of total production prior to Norton-Griffiths' raid. The extent of the destruction caused by British forces had a detrimental effect on Germany's ability to make war. Furthermore, the meticulous and relentless effort to disable the Romanian oil industry, and, consequently, the scale of the recovery undertaking exemplified the importance of oil by the British and German governments.

In the immediate post-war period, Romania, financially distressed from a disastrous war against the Central Powers of Germany and Austria-Hungary, was facing extortion. A key oil producer for Germany before the war, Romania was facing a reconstitution of her oil industry in favour of Allied interests in exchange for what was, effectively, a bailout. After a German annexation, the Romanian oil industry was controlled by German financiers, companies and banks acquiring the rights over oil rich lands and associated assets. By early 1919, however, Britain and France moved swiftly to request that all German presence from Romania's oil industry be removed and replaced by interests more favourable to doing business with London and Paris. A series of loans desperately needed by the new Romanian government was contingent on this particular prerequisite. The Romanian case, particularly its aftermath, is a good example of two powers vying over the resources of a lesser power and the impact – notwithstanding the fact that the circumstances were far more precarious – an aggressive policy had on the economy and environment of a smaller state.

Indeed, German interest in the potential of oil manifested in many different forms. From the Treaty of Brest-Litovsk in 1917 promising 75% of all the oil produced from the oil rich area of Baku to be exported to Germany and her allies, to Deutsche Bank investing heavily in companies and establishing banks such as the National Bank of Turkey which had access to concessions that would be important for the search and exploitation of oil in regions of the Ottoman Empire, Berlin pushed ahead to not only aggravate

2. Daniel Yergin, *The Prize*, (New York, 1991), p. 181.

British pursuits in oil but also to see where she could gain any advantage over her adversary. The records held at the *Bundesarchiv* went as far as to show efforts to secure oil from the U.S. and Mexico; such was the desperation of the German government.

Likewise, Britain, and to some degree France, began to systematically decide which companies would be given access to oilfields in the various mandated territories and protectorates. With areas in North Africa being identified as potential zones for exploiting oil, the British were content with allowing the French government to oversee that operation. The same was afforded to the British who now sought to administer Persia (present day Iran), Mesopotamia (Iraq) and Palestine. Even the European states were not free from British and French interests. The re-establishment of the state of Poland also saw British and French influence pique as the borders, yet to be determined, would possibly include a rich oil basin from the former Austro-Hungarian Empire. How this would be determined was of particular interest to London and Paris.

In the post-war Weimar Republic of Germany, the lessons learned from technological inferiority were felt the hardest. Many of the documents I discovered related to the efforts made by numerous German ministers trying to secure new sources of oil for a dysfunctional economy. Even within Germany itself, industrialists and operators at refineries wrote to the economic minister in Berlin asking if and when they would be able to manufacture goods or refine what oil they still had access to. The smothering effect of the Entente economic blockade and strict terms of the Armistice ensured Germany would not only be unable to produce material, but also utilise resources associated with war manufacturing. To be sure, while Germany may have adopted an apathetic opinion over the viability of oil prior to the war – especially as she did not have any domestic supplies to exploit – Berlin's attitude reflected a remarkable reversal of policy after it.

In summary, the research trip to Germany provided a substantial number of documents and records on how the oil industry cooperated and collaborated with the German government, before, during and after the First World War. When compared to the same practice by the British government, there is a clear transition from indirect and sometimes direct support, to becoming a primary objective for existential reasons. The British government identified this fact much sooner than the German government and this is reflected in its policies and actions towards oil-prolific regions and the companies seeking to exploit them. However, the German government unsatisfied by the terms of the Versailles Treaty began a desperate struggle to ensure ease of access and abundant supply as it sought to rebuild its economy after the war and prepare for the next European conflict.

This project will shed a light on the history of the relationship between the oil industry and powerful government. The modes and practices that would seem questionable even for the time they were carried out would lay down a framework and precedent for unsavoury policies and action for the decades to come. As oil became intrinsically linked to not only economic prosperity, but also technological superiority, its acquisition necessitated policies to ensure a plentiful supply because without it, powerful states would see their reach diminish and status decline. Furthermore, with the rise of large corporations and the free market principle, governments came to rely upon these suppliers to maintain their growing economies – as more recent times have demonstrated, the movement and sale of oil now possesses the ability to influence the global economy. Through this project and the support offered by the Ken Saro-Wiwa bursary, my work offers some insight into the origins of this relationship between government policy and the emergence of what would become the dominant source of energy in the twentieth century. Moreover, I would argue, the beginning of this great rivalry and scramble to analyse and discover the importance of oil plays a key role in understanding our modern world.

A dying village

MAJELLA MCCARRON

Sr. Majella. Maynooth University Ken Saro-Wiwa Archive.

No piercing siren to rise you,
No flaring light to guide you,
No nearby neighbour's love to reach you,
Dying village.

No caring friend to conceal you,
No towering soldier to defend you,
No rushing firemen to quench you,
Dying village.

No speeding ambulance to ferry you,
No humble priest to bury you,
No Red Cross pennant to fly for you,
Dying village.

No urgent phone to ring for you,
No loud-pitched radio to plead for you,

News is blocked in fear of you,
Dying village.

No strong one comes to hold you,
As children are torn from you,
A stranger's voice to wail for you,
Dying village.

Dawn comes late for you,
Vultures chuckle over you,
Our deepest human shame is you,
Dying village.

Too few prophets spoke for you,
Years of scribes and Pharisees denied you,
Evil powers abandoned and beggared you,
Dying village.

The world turned its back on you,
May God himself be good to you,
And hope renew in you,
Dying village.

KEN SARO-WIWA'S LETTERS
TO SISTER MAJELLA

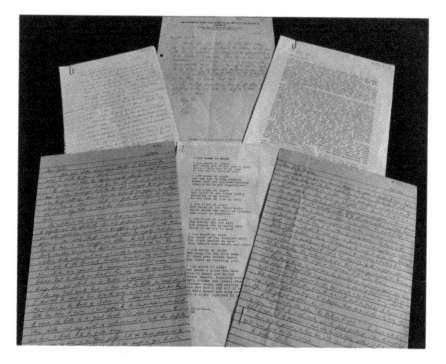

Selection of letters and poems from the Ken Saro-Wiwa Archives.

Note from the Editors

WITH THE EXCEPTION of the letters dated 8/7/1995 and 16/9/1995 which were typed, the letters transcribed in this volume were handwritten. They inevitably contain idiosyncrasies such as contractions, distinctive choices of spellings (e.g. "gaoler" rather than "jailer"), and occasional slips of the pen. In places also, the format of a letter has been complicated by Saro-Wiwa's emendations including, for instance, notes in the white space or margins.

The editors have aimed here to transcribe what was written as it was written, or, in other words, to reproduce in typescript as accurate a version of each letter as possible. In doing so, it has sometimes been necessary to make strategic decisions to provide a clear text. For clarity, we list these decisions here. The reader should note that the Maynooth University Library is eager to facilitate any person who, for research purposes, seeks access to the original letters held in the archive.

When a note or other emendation was intended by Saro-Wiwa to be integrated into the text, we have done so. Where the appropriate place for a note is less obvious, it has been included in full in the footnotes instead.

At Majella McCarron's request, five sentences in total have been deleted from the entire text of the letters to protect the personal information of tangential individuals. These deletions are indicated within the text by three dots inside square brackets. Since Saro-Wiwa's own deletions (or cancellations) are frequently illegible, they are not signalled here.

Throughout the typescript, we have retained Saro-Wiwa's contractions, underscores and choice of spelling. In the cases of contractions, we have added editorial emendations within the text to clarify their meaning. These are signalled by square brackets and include, for instance, the full spellings of names or words and the full forms of acronyms. Other authorial emendations are also signalled by square brackets.

To facilitate fluent reading, we have made a small number of silent emendations where slips of the pen resulted in the occasional misspelling of ordinary words or obvious minor errors in punctuation.

October 20, 1993

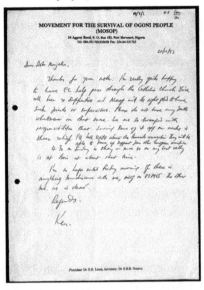

Letter from Ken Saro-Wiwa to Sr. Majella McCarron. Maynooth University Ken Saro-Wiwa Archive

20/10/93

Dear Sr. Majella,[1]

    Thanks for your note. I'm really quite happy to have EC [European Commission] help pass through the Catholic Church.[2] You've all been so sup-

---

1. McCarron was working as a lecturer at the University of Lagos when asked by the Vatican to act as an observer for the Africa Europe Faith and Justice Network (AEFJN). Her activities with the AEFJN led her to make Saro-Wiwa's acquaintance and when the university personnel were on strike, McCarron often spent time at his business office in Lagos to learn from him about the Ogoni situation. The AEFJN describe themselves as "a Faith-based International Network present in Africa and in Europe, established in 1988. [They] promotes economic justice between the European Union and sub-Saharan Africa so that the poor of Africa may look toward a better future".

2. When Ogoni villages were destroyed in September 1993, McCarron made a successful

portive and MOSOP will be right glad to have such friends or supervisors. Please do not have any doubts whatsoever on that score. We are so swamped with responsibilities that having some of it off our necks is sheer relief. I'll tell UNPO [Unrepresented Nations and Peoples Organization][3] about the Brussels connection. They will be able to drum up support from other European countries.

4.30 on Sunday is okay—more so as my last rally is at Bori at about that time.

I'm in Lagos until Friday morning. If there is anything to discuss with me, ring on [telephone number provided]. The other tel. no. is dead.

Regards,
Ken.

application for EC humanitarian aid to provide village relief. It was an EC requirement that a European non-governmental organisation would co-operate with a local non-governmental organization to manage and deliver the funds. McCarron contacted Niall Tobin in Trócaire, an Irish non-governmental development agency, to secure their support.

3. An international democratic and non-violent organization dedicated to promoting the rights of populations who are not represented at major international fora, such as the UN. Its membership includes indigenous peoples, minority populations and the populations of unrecognized or occupied territories.

December 1, 1993

Maynooth University Ken Saro-Wiwa Archive.

1/12/93

Sr. Majella,

I read your notes with great pleasure. I'm sorry to hear about the illness of your mum. I hope she recovers. Please keep putting your thoughts on paper. Who knows how we can use them in future. The Ogoni story will have to be told!

Can I please prevail on you to send the attached envelope through your special courier to London? I'll be most grateful.[1]

There's a lot for us to discuss. I'm in Lagos again next week Wednesday. We must please meet either Wednesday evening or early Thursday. I expect to return to Port Harcourt on Friday.

God's blessing,

Ken Saro-Wiwa.

---

1. McCarron often asked friends who were travelling to post letters from outside Nigeria as the Nigerian postal system was sometimes unreliable.

July 13, 1994

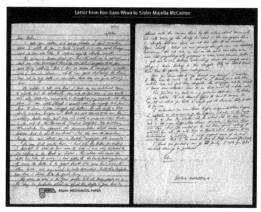

Maynooth University Ken Saro-Wiwa Archive.

13/7/94

Dear Sr.,

I got your letter, and many thanks. I don't remember what I wrote to you, I think I wrote in a very great hurry and so was not able to express myself clearly, maybe.

Of course, I know that you are all with me in spirit and am very encouraged thereby. Nor did I imagine that you were doing nothing. But I have no access to newspapers or radio, and I was in chains—which was quite depressing. The chains now sit on my table, a reminder that they can go on at any time.

My condition is not very bad. I have an air-conditioned room to myself, and the electricity has only failed once. I can write and only yesterday succeeded in smuggling my computer into this place. I can cook (though I cannot cook) for myself and from time to time, I can smuggle out letters.[1] The only thing is that family members, lawyers and doctor are not allowed to see me. The military doctor came just once and wrote a recommendation that I be sent to

---

1. Family members were required to supply food as was the norm in Nigerian prisons. The letters were smuggled in and out of the detention cell in food baskets.

the University Teaching Hospital. The Military Administrator [Lt.-Col. Duada Musa Komo] has ignored the recommendation which makes me believe that he wants me dead. I've also seen the scurrilous things he's said about me in *Quality* magazine. It's annoying!

In my first month here, I had only the Bible for reading. I decided to read it from cover to cover. I was very disturbed by the violence in the book of Joshua. The soldier who owned the Bible has taken it away. I had gotten to the lamentations of Jeremiah. Of course, the Bible is a great book. I've since had access to other books and my mind is well-nourished. With the computer now here, I think I'll be quite busy.

My worry, as ever, is the Ogoni people. With all MOSOP people out of the way, the protection which we offered the people is gone.[2] But I'm pleased with the concern shown by the international community and with some of the things I've read in the newspapers here. I strongly believe that we will be able to re-create Ogoni society. What we are passing through now was absolutely necessary. It's not even as bad as the civil war when we were not psychologically prepared & were mere cannon fodder! I get one or two letters indicating that the people remain strong in their belief in the struggle. Pity we didn't teach them how to operate from underground!

2. Following the arrests and violence, many MOSOP members went into hiding. Some including the author's brother, Owens Wiwa, had fled the country; a number received shelter at a refugee camp in Benin.

Relief workers rebuild Ogoni villages with the EC emergency funding that was released to the Diocese of Port Harcourt through Trócaire in Ireland. Maynooth University Ken Saro-Wiwa Archive.

Some Ogoni inhabitants survey the impact of violence along the waterfront at Port Harcourt, 5th January 1994. Photograph by Sr. Majella McCarron. Maynooth University Ken Saro-Wiwa Archive.

Have you been able to go through my essays? I wonder if they can be made into a book or I should just take the central ones and put them alongside the newspaper articles which I have now collected. Seems to me that would be a better idea.

I've finished my book on my previous detention. I'm looking for an editor. I'll ask my office to send it to you. You can tell me what you think & maybe get the American lady to edit it for me.[3]

I understand you now have bigger responsibilities. Congrats. I expect a denouement of the Nigerian situation in the next month or two.[4] My situation will be solved with it, but there are a lot more difficulties ahead. However, I believe that I've done what God wanted me to, and have spared nothing to achieve his will. He will have to decide what happens to me. Incarceration is nothing. I must expect more of it, and even death. But I do want to live to help re-create Ogoni society.

I thank you for taking care of my family. I wish you God's abundant blessings & protection.

Ken.

3. The book in question is *A Month and a Day: A Detention Diary* first published by Penguin books in 1995. The "American lady" to whom Saro-Wiwa refers is poet Lynn Chukura who came to Nigeria from Philadelphia to lecture in Lagos and Legon. As it turned out, she did not edit the manuscript.
4. On 12 June 1993, there was an attempt to restore civilian government. Abiola was elected but prevented by the military regime from taking office. Sani Abacha's dictatorship continued until he died in 1998. Abiola died the same year.

Undated letter

Dear Sr. Majella,[1]

Greetings in God's name. Well, you know the whole story. They are getting closer to me—Shell and the Nigerian establishment that is. I'm not particularly protected, although I have great faith in God, in the justness of my cause & in the belief in eventual VICTORY. But the pain which we all have to endure! Would to God it had been lighter!

My current detention is sheer torture. I'm a private prisoner of the Lt-Col Komo[2] and his Internal Security Task Force. This is a lawless situation. I'm not being held under Decree 2 for all I know, and if I were being held under the Criminal Code over the homicide of the 4 Ogoni men, I should be in the hands of the Police. [UNHRC have a mandate on the condition of prisoners. Can the rep in Lagos do something about me?][3] Now here I am, in a private house, denied access to lawyer, doctor, family, other visitors and not allowed to have the special diet which I have been on. I am not allowed to read newspapers, listen to radio or read books. It's mental torture. The living condition is okay—there is electricity & air conditioning, but I'm alone with two armed guards, 24 hours a day. I've asked Bishop Makozi [Catholic Bishop of the Port Harcourt Diocese] to intervene with the Governor so I'm properly fed & taken to hospital. No dice. You should see me. I've lost weight! For the first 10 days here, I was on bread, water & bananas alone. But I'm in good spirit, undaunted, as convinced of my cause as ever. My real worry is the devastation of Ogoni villages, the destabilization of the area & the harassment & killing of the people. With MOSOP Steering Committee members on the run or under arrest, the Ogoni are not protected at all. And the international

---

1. The following letter was not dated but McCarron believes it was written in June or July 1994. On 21st May 1994, four Ogoni elders—Albert Badey, Edward Kobani, Samuel N. Orage and Theophilis B. Orage—were murdered at Giokoo in Ogoniland. Saro-Wiwa and three other MOSOP activists were accused of conspiring in the murders, arrested on 22nd May and detained without charge.
2. Colonel Duada Musa Komo, Military Administrator of the Internal Security Task Force which was set up to deal with the Ogoni campaign. For more information, see Timothy Hunt, *The Politics of Bones: Dr Owens Wiwa and the Struggle for Nigeria's Oil* (Toronto: McCleland and Stewart, 2005).
3. The bracketed sentences were added in a column along the side of the paragraph.

scene is quiet, taken up with Abiola.[4] Only Divine intervention can help the Ogoni.

I'm not worried for myself. When I undertook to confront Shell & the Nigerian establishment, I signed my death warrant, so to speak. At 52, I think I've served my time and, come to face it, I've lived a charmed life. A few more books, maybe, & the opportunity to assist others would have been welcome. But it's okay. Of course, I & MOSOP had nothing to do with the death of the 4 gentlemen. We are struggling for justice, not for power and, in any case, they were of little consequence in a highly mobilized and conscious Ogoni population. They were no threat in any way at all. Komo has just succeeded in masking the government's role in the unfortunate & brutal deaths. And the Orages[5] were my in-laws. My children are cousins of the Orages. Elizabeth is from Bane & is the elder Sr. [sister] of my estranged wife, Maria. And I did have a lot in common with Edward Kobani.[6] We continued to discuss & chat even in recent times. We always got together again after he'd have strayed to his heart's content.

Well, Sr., I hope you do get this letter. I hope I'll get another opportunity to write you. I'm spending my time writing short stories—I lock my door & do not allow my gaolers [jailers] to see me at it.

If we meet again, we'll smile. Till then, it's good luck & God bless you.[7]

4. Moshood Kashimawo Olawale Abiola (M.K.O. Abiola) ran for the office of President in June 1993. He is widely presumed to have won the election which was annulled by the presiding military president, General Ibrahim Babangida, before an official result was announced. A political crisis ensued, allowing General Sani Abacha to seize power some months later.
5. A reference to Samuel N. Orage and Theophilis B. Orage, two of the four Ogoni chiefs who were murdered at Giokoo in Ogoniland on 21st May 1994.
6. Another of the four Ogoni chiefs who were murdered on 21st May 1994.
7. This letter is unsigned.

July 24, 1994

24/7/94

Dear Sr. Majella,

Don't say you are leaving Nigeria for good. I tended to read that in your letter and it left a hole in my heart. Thanks for the letter anyway. It cheered me a lot. I'm in good spirit and quite content to leave myself in the hands of almighty God. Somehow, I'm finding a lot of activity–reading and writing. I've now completed a volume of short stories. I've actually written five of the stories before now. I've done 5 more & gotten a book [probably *A Kind of Festival and Other Stories*]. I start on re-writing the novel I lost in 1992 [*Lemona's Tale*] at the end of next week. It's rather nice that I have unfailing electricity here & the room is comfortable. I'm learning to cook–I am allowed to cook my own food & NEPA [National Electric Power Authority] supplies me with fresh fish regularly. I think I'm beginning to LIKE it if you can believe me. The only problems: lack of access to relatives & doctor & lawyer.

I had my aged parents (Dad 90, Mum 73) go to newspaper offices in PH [Port Harcourt] to embarrass the authorities a bit. Very mean of me, isn't it. So you may see them in the mags & newspapers before you leave.

I'm trying to be as inventive as possible & keeping my spirit up in that way. It's good that my conscience is clear & that I realise things could be worse. I'm worried for the Ogoni people & for Hauwa[1] & my children. But somehow, I think they have to live with such buffeting. So long as I am fighting Shell, so long will I suffer. But that is the only way the Ogoni will get out of their bind.

Have you seen the film *The Drilling Fields*?[2] You must ask for a copy.

I wish I could really see the final product. I saw the rough cut before I returned to Nigeria. I actually have a copy of that in P.H. [Port Harcourt] or Lagos.

Have a good journey & God be with you. Keep me in your prayers & don't ever forget the Ogoni people.

---

1. Hauwa Maidugu, Saro-Wiwa's partner at the time of his detention and mother of his youngest son, Kwame.
2. A 50-minute documentary film examining the activities of Shell and other oil and gas corporations in the Niger Delta. It was directed by Glen Ellis, produced by Poonam Sharma for Catma Films and first shown on Channel 4 London on 23rd May 1994.

A million thanks.
Ken.

July 30, 1994

30/7/94

Dear Sr. Majella,

I got your papers this morning & was I excited! You probably have not got the letter I sent through Barika.[1] Should be waiting for you in Lagos.

I will miss you in the year you are going to be away! Your letter was the first really good intimation I had of the goings on in Ogoni. I've always realized that we have not prepared the generality of the Ogoni people adequately for the struggle. The march of January 3 should have come after the organizations were properly in place.[2] But it was difficult to organize anybody or anything in Ogoni without such a bold public statement. It was like grabbing the place by the scruff of the neck and thrusting it forward.

In spite of this, I remain optimistic. The military government should collapse shortly after you leave Nigeria. That will open another chapter. The need now is to get the military out of the place. I hope that Ogoni is used to cudgel the military culprits who may do a deal with Abiola & so escape the charge of treason. But they must not escape a charge of genocide which they probably will not be thinking about. As Claude Ake[3] said, we need an international inquiry to go into it.[4]

Aid cannot be properly channelled into the area until the military are gone & we begin to re-organize the people. What happens during your absence?

I'm including a letter for the Bodyshop[5] & UNPO, along with their fax and telephone numbers.

I write fiction, think about Ogoni & pray to God. I'm also trying to direct

---

1. Probably Barika Idamkue, a fellow activist and member of MOSOP.
2. The author is probably referring to the first mass public rally held in Ogoni on 4th January 1993. 1993 was the UN-declared "Year of Indigenous People".
3. A prominent Nigerian political scientist.
4. Dozens of people were killed when Ogoni villages were destroyed in violence that took place during September 1993.
5. A British-based company, founded by Anita Roddick in 1976, which works on the ethical sourcing of beauty products. Saro-Wiwa had met with Gavin Grant who was head of corporate communications and public affairs for the Body Shop International in 1992.

Ogoni affairs from here. It's not easy, but I'm doing my best. The scattering of the MOSOP committee is sad. But we have to learn to operate from the underground. We need training in too many areas! I'm happy you met Ben Naanen.[6] He is a fine fellow.[...]God bless you & travel safely.

Ken.

My telephone no. in London is [number provided]. Speak to my daughter Zina.

My son's number is [number and address provided]

I haven't heard from the children & that's worrying. I and my wife [Maria], as you probably know, have not been at one. And the family has been wracked by illness & tragedy[...]My first son underwent heart surgery last year[7] & the Orages who were murdered were in-laws. Sam Orage brought up my wife (he's married to her elder Sr. [sister]). So the tragedy grows. I've tried to keep this personal aspect away so that it does not interfere with my public responsibilities. The girls (twins) are fine though, and Zina is especially a delight & very clear where she wants to go to. She needs guidance, though, & I've not been able to provide that.

P.S. By the way, I was sent here when it was found that I was sending messages from the Bori Camp Guardroom[8] through the food basket sent by my family. The military introduced Ogoni moles into the Guardroom. We knew them well. When one of them told on me, it was time to move me out. It has proved to be a blessing. At last, I'm writing & am able to think clearly in good conditions. And, sorry to say, am still communicating with the outside world, in the absence of the food baskets!

Ken.

By the way, I don't consider Birabi & Leton a problem as such.[9] There is no

---

6. A fellow activist and member of MOSOP. He was Professor of History at the University of Port Harcourt and has collaborated with the UN Research Institute for Social Development on a project investigating Identity, Power and the Rights of Indigenous Peoples 2006-08. He is now Acting Head of MOSOP.

7. Following the death of Saro-Wiwa's son Tedum at the age of fourteen, the family were tested for heart problems. Ken Jr. was required to undergo heart surgery.

8. An Internal Security Task Force centre in Bori, Ogoniland. The guardroom was similar to a cell in a police station where people may be held before being charged. Bori is the main urban centre in Ogoniland.

9. Bennet Birabi and Garrick Barile Leton, both members of the conservative Ogoni elite. Leton had been a former President of MOSOP but testified against Saro-Wiwa

negotiating with them because they do not even want to negotiate & they will not accept the democratic decisions of the majority. Really, all of it is based on jealousy & frustration. The people will have to decide what to do about them. Invariably, I find that the younger chaps want to fight them to a standstill. They matter only when the authorities side with them. So let's see how national politics go. If a Sovereign National Conference gets held & the Ogoni obtain self-determination, then the political education the majority have received will reduce dissenters to an insignificant opposition, incapable of frustrating the people.[10] MOSOP have 90% support in Ogoni And that's okay, by any standards. There was a beautiful essay in *Daily Sunray*[11] of Thursday 28.

Did you see it?

Ken

at the military trial. According to a report by Human Rights Watch, he "alleged that Saro-Wiwa had sought complete control of the organization and encouraged his supporters to employ 'militant tactics'." But, in fact, the report continues, Saro-Wiwa had "repeatedly avowed his commitment to non-violence" and "criticized the actions of violent Ogoni splinter groups". See <http://www.unhcr.org/refworld/docid/3ae6a7d8c.html>

10. There were plans afoot for a conference to take place in Abuja in 1994 on behalf of all of the ethnic nations of Nigeria but they did not materialize.
11. A Nigerian newspaper based in Port Harcourt.

August 15, 1994

15/8/94

12 noon

Dear Sr.,

A million thanks for your letters. They are so entertaining, so encouraging, & they give me those intimate details of my family which no one gives. God bless you & keep you for us!

And you cannot imagine the relief I got from the money you sent. I was at my wits end. My resources are not limitless. They've begun to flounder. I've spent ever so much on the Ogoni. Runs into millions of naira. And now, something will have to happen. Otherwise I'll begin to sell my property. Just too bad for the youngest Saro-Wiwa [Kwame] whose birth am so happy about! I know that my daughters are the ones who will perpetuate my memory (apart from the books, of course,) but I have a sense of balance & am not willing to increase the number of disadvantaged in the world–women being so, most unhappily & something we must sort out too! More battles to fight!

So thanks for the cash. God bless the giver(s).

I don't know if you did see *Sunray*. The girl, Meesnan Akekue whom I trained–she worked with me for 2 years & I gave her a bursary for an MA at Uniport [University of Port Harcourt]. She has been writing some good pieces on the soldiers in Ogoni villages. Which solves your problem since I could not well know what's going on from here.

After my piece in the *Sunday Guardian* [the *Guardian*, Nigeria] of 31st July–did you see it?–the Administrator had my computer and radio confiscated. I was happy the piece rattled them so much. I miss the computer, but I feel really good that they've been so miffed!

The newspapers and magazines have been most supportive & am really happy about that too. The power of print! As a result of Rogerson Ake's article (a second, more devastating one is due soon), the beating of Nick Ashton-Jones[1] & probably my article, the ISTF [Internal Security Task Force] & Major

---

1. British environmentalist and co-author with Susie Arnott and Oronto Douglas of *The Human Eco-Systems of the Niger Delta* (1998), a handbook of Environmental Rights Action (ERA). He conducted an Environmental Impact Assessment (EIA) in the Niger

Okuntimo[2] have been queried by Army HQ. I hope he gets it this time round. The Sadist!

There are rumours that we might get charged to court this week & remanded in prison custody. This will be much better that the present brutal detention where I remain incommunicado.

I hope you did get the letter of recommendation to UNPO & the Bodyshop. I'll be expecting you to carry the Ogoni banner in Europe in the next week.[3]

It's going to be a difficult battle, this Ogoni struggle. But I've not had any doubts at all of success. I've recently gotten even more confident as incarceration has drawn me nearer to God. He works in a mysterious way, his wonders to perform. What doubt can there now be of the Ogoni angst in Nigeria? I've even begun to think of Ogoni asking for a UN Protectorate status so that we can be protected against Shell & the other Nigerian groups. A line I may pursue if only to scare Nigerian rulers.

I hope that Ogoni presence in Geneva this month will yield some attention from the UN Human Rights Commission. That should be good for us.

I expect this letter to get to you in Europe. Please take care of yourself and do contact my eldest boy. He's a fine fellow—well educated but in the mould of an English gentleman (pity). He's slow & deliberate & cannot really understand what the barbarians are doing down here. He'd like to know how else he can help. International Pen[4] might find you of value before literary audiences so contact them.

I'm having to write a dozen or more notes & letters to my staff etc. Wherever you may be, please remember that there's someone here who appreciates your goodness, your works. I'm sure my daughter Singto will always remember you too. She's really intelligent. I'm not surprised she was watching you so carefully. Three or four years back, she wrote a beautiful 20 page book on computer on her dog. It's a literary masterpiece which I hope to

Delta area for the Africa Development Bank and Federal Environmental Protection Agency in 1994.
2. Major Paul Okuntimo of Rivers State Internal Security Task Force.
3. McCarron was due to return to Ireland the day after this letter was written. She had not renewed her contract at the University of Lagos where she had been teaching for 13 years.
4. International PEN is a global community of writers who promote freedom of expression in literature.

publish. She's observant & thinks for herself. I'm proud of her. As of her Sr. [sister] Adele who lives in England. She's extremely gentle and sweet—not as clever as Singto but gentility is a great asset. With Zina (the first of the twin girls), I have a real team of capable women, if they do not meet & get enslaved by some mean men!

Keep writing Sr. M & God bless you a million times.

Sincerely,

Ken

September 16, 1994

16/9/94

Dear Sr. Majella,

It's hard to think you've been away for a whole month! How fast time does fly! I saw your picture holding my young Kwame whom I'm still to see. Thanks ever so much for caring, and for the money you sent us. It came in really useful. I wonder too if you're behind The Bodyshop asking after our 1994 Budget. If they could help in any way at all, it would be great relief for me. I was thinking of selling some of my property to keep the struggle going!

In the month since you left, I see the situation in N. Ireland has improved tremendously. The possibility of peace is so comforting. I hope it happens. 25 years is a long time to be fighting, surely. God grant that it works.

Nigeria has progressively gone down the drains to its worst possible nadir. With all sensible newspapers barred, a lot of people in detention & laws which establish that the dictatorship cannot be challenged in court, we are in real trouble, to say the least. I don't see Abacha[1] lasting much longer anyway.

I'm still here (118 days today) and there's no sign that I'll be out soon. No matter. I'm writing and reading a lot and making good use of the time. One of my poems was published in the *Independent* of London on 8th September. And Professor Ake & Wole Soyinka[2] have done excellent essays on Ogoni. That by Wole Soyinka appeared in the *New York Times* and probably sent the Nigerian Ambassador to the U.S to Ogoni. He came here to see me; unfortunately, we could not really discuss meaningfully because the military upset me by remaining with us. I also wore a MOSOP singlet to ensure that they did not get a photo opportunity! They had a video camera along & the *Sunray* photographer smuggled himself into the team. Later, the *Sunray* films were confiscated, the editors of the newspaper manhandled, and the MOSOP singlet seized from me! But I had made the point.

The published essays by Ake, Soyinka & myself have lead to an upsurge of

---

1. General Sani Abacha, Military Dictator of Nigeria 1993-1998.
2. Major Nigerian author, the first African winner of the Nobel Prize for Literature in 1986.

confidence in Ogoni. I now receive a stream of encouraging letters from the activists, including those of them who are underground. The sense I have is that they will "fight" to the end. Barika is off to the U.S., after his stay in Geneva, to speak on the Ogoni issue & Professor Ake tells me the Europeans & Americans are doing a lot on the Ogoni issue.

The British chapter of International Pen made me an honorary member & they have written to ask how they can get me out of jail. I asked them to appeal to Shell & to use the British media as much as possible to keep my case on the front burner.

The government keeps saying we are to be charged to court "soon". But they've been saying that for about 2 months. I doubt that their investigations led to much, but they are probably looking for the judges who will agree to do their bidding. I've seized upon their statement to suggest that the Ogoni people set up a "Saro-Wiwa Legal Defence Fund" as a rallying point. I'm told the launch of the same will take place at the Roman Catholic Church at Garara on the 18th.

Abacha's Constitutional Conference[3] is thinking of tinkering with the creation of nine states—and a Forum in Rivers State recently suggested the split of the state into 4, including an "Ogoni" State. I hear Shell are pressing for the creation of "Ogoni State" which was one of the things I told Emeka Achebe [Executive Director for Shell][4] unofficially that they might do if they wanted peace with the Ogoni people. I don't have faith in Abacha's regime anyway and I doubt that much will come of his Conference. After the collapse of the oil strike, there has been quite some despondency in pro-democracy circles, but everyone still believes that the desperation of the regime signals the beginning of the end. I can't wait for that end.

All in all, I'm in high spirits and my time is well used, the physical conditions have been quite good. Had my computer not been confiscated, I should have written a lot more than the 5 or 6 books I've now either finished or com-

3. The conference referred to as the Sovereign National Conference in the final addendum to the letter dated 30/7/94. See MU/PP7/6.
4. Executive Director for Shell from 1981-1996. He was General Manager of Shell's operations in Nigeria at the time of Saro-Wiwa's detention and later appointed Senior Corporate Adviser to Shell in London. "Emeka" is a short form of "Nnaemeka" and Saro-Wiwa uses both iterations alternatively in his letters here.

pleted anew. And somehow, I remain immensely confident that the Ogoni cause will be won.

I hope that you meet your family in good health and that you are resting as you should.

Please keep in touch.

Sincerely,

Ken.

October 1, 1994

01/10/94

Dear Sr. Majella,

Thanks a lot for your letter starting from 3rd September & the only communication I've got from you since you left. I have, on my part, sent two letters and doubt now that you got any of them. I'll check what has been happening at the Lagos end, where there has been quite some confusion and lack of funds has stymied movement. Fundwise, I got a respite from the banks which have seized my funds invested with them—just one bank. I've got money to keep going for two months.

Keeping MOSOP operational has been quite a problem. With all the Steering Committee members in jail or declared wanted by the police, we have been in trouble. People like my brother who was helping have themselves become a liability. Those who were in jobs have had their salaries stopped. They, too, have become my responsibility. And keeping things moving in present-day Nigeria has become very expensive—and one is living on savings—earning nothing new for now. I did mention in my last letter that the Bodyshop had asked for our budget. If they can chip in something, it will help a great deal.

You probably know that one of my aims has been to take the Ogoni people on a journey. Even what is happening now is, and please don't think me sadistic, helpful. For one, they are able to see me battling from prison— from the very jaws of the lion. A number of them have stuck it out in Ogoni and are still able to work in cells. And there are those who went off to Lagos and have done marvellous work with the Press. The activists write me, and from them, I have a sense that the Ogoni people are holding out bravely. They are not fighting—because I did not even prepare them for physical combat—but they are holding out psychologically. And that, in spite of massive government propaganda, aided by renegades like Birabi And Leton.[1]

However, we have won the propaganda war. I hope that you have seen the writings of Professor Ake and Wole Soyinka—the latter appearing in the *New York Times*. Locally, the support of the non-government press has been

---

1. See n. 26 with the letter dated 30/7/94

tremendous. And Yoruba leaders meeting on August 31 sent solidarity messages to the Ogoni and called for my release.

The Rivers State Government have put out films which I have not seen but they are recognized for what they are—propaganda—and in any case, it's always difficult to sell a bad case. More so when the opponent has had a head-on start. Nor can they match us internationally.

I thank you for your thoughts on the *Ogoni Review*. I have been wanting to keep it going. Funds have been the problem. If you can raise money abroad for it, we can do all the work, including printing here. Barika has been to Switzerland, and then went off to the U.S. I have not heard of him since then. I have, all the same, had a new employee who's working on copies. I'm sending back copies with this letter in the hope that it will help you raise funds. Three hundred pounds (£300.00) sterling a month should be enough to produce 1500-2000 copies monthly & mail them free to the right people and organizations. If such funding becomes available, all you need do is pay it into my a/c no. [number provided] at [name and London address of bank provided]. I'll then arrange to repatriate the funds through the unofficial market which will make the naira equivalent even better.

I'm happy that Bodyshop were able to stop the bad publicity which I'm sure was engineered by Shell. They may be made to realize that what we are doing in Ogoni helps to expose Shell as well and they should lend us support. It's a shame that MOSOP have not been able to get help except from the World Council of Churches (cash) and UNPO (kind).

You will have known that the strike in Lagos was called off after 2 months. It was a great effort, but the British Govt. helped with it. The oil companies too. These organizations find it easier to exploit Nigeria through the military dictatorships. Predictably, Abacha has gone on a spree, trying to prove that he can out Amin Idi Amin. But I expect that he will fail, ultimately.

Yes, I do have a radio. Two have been seized from me, but I've got a third. If they seize that, I'll get another. It's the only way I can keep up with events. I get the newspapers too, but all the good ones—*Guardian, Punch* & *Concord*—are proscribed. But the radio has been most useful. Since I have time on my hands, I'm able to follow a wide variety of programmes. I get overseas magazines once in a while—through Prof. Ake—and I do have access to the novels in my library. I've written quite some—novels, Mr. B books & the collection of short stories. At least the first draft is finished and that's some comfort. My time is well used. Getting all these things in has meant paying money out to my guards—quite a sum of money, Nigeria being Nigeria. But that's okay. I can put it down to "business expense". Freedom can be quite expensive or cheap depending on how you look at it. To those who have freedom, it's cheap; those of us who lack it, pay a lot to get just a bit.

Hauwa was here for a fortnight or so but could not get to see me so I was not able to see my son, Kwame. That the authorities refused to allow it shows the depth of their wickedness. But in Ogoni custom, to refuse a child the favour of seeing his parent is godlessness and a crime. So I'm expecting the Ogoni God to punish the criminals. Once when Edward Kobani, now deceased, treated me in a dastardly fashion, my father predicted that he would not die well. That duly came to pass. My dad did see and bless the child along with my mother. He gave them a lot of happiness, I hear, and am satisfied. I've also had a guard here (a soldier) who saw my parents in good health.

I'm pleased that you met Father Mashevran.[2] I don't think I've ever been "street-wise". Bull-headed, yes. You have to be to take on Shell and the cabal that rules Nigeria.

The advice for Ogoni people not to co-operate with the military came from you, of course. Did you think I wasn't hearing from you?

And now to yourself. I hope that your medicals prove you fit. And that you are well, and happy. I long to see you back in Nigeria, helping, among others, to guide the Ogoni people through the wilderness. You don't know what help you have been to us, and to me personally, intellectually. God grant that you do return to us. I'm counting the days. I may still be in detention when you come back. But I'm not worried about that. Since the physical conditions are not bad, I'm keeping myself mentally busy and doing a lot of those things which I may not have done as a free man. Has it not been said that God moves in a mysterious way/His wonders to perform? When I think how I came to be here, and to succeed in internationalizing the Ogoni issue on a slim budget, I cannot but see God's fingers in it all. And evidence is now getting out to the effect that I might have been the one to be assassinated on 21st May but that what had been planned for me went askew, thanks to God. Not that death would have mattered to me. It would have carried more harm to those still alive. However, I do want to take the Ogoni people as far on the journey to re-vitalization as is possible—until other leaders are bred.

2. Saro-Wiwa's handwriting is indistinct here and it has not been possible to verify the spelling of this name.

My brother, Owens,[3] is lying low in Lagos. But he's doing a tremendous job with the Press and the embassies. He's very clear and conscientious, thank God.

I've dealt with all the issues you raised in your letter and as I sent a letter earlier, I've probably gone over some of the issues I raised in that letter.

Let me end by wishing you and yours the very best and God's abundant blessings.

Ken

---

3. Dr Owens Wiwa, medical doctor, a fellow member of MOSOP and global human rights activist. He is also currently Nigerian Country Director of the Clinton Foundation.

October 11, 1994

11/10/94

Dear Sr. Majella,

Welcome back to Nigeria and to Ogoni.[1] I hope you did receive all my letters. I do not appear to have been so lucky. Anyway, you can write me all you want. You might even try to see me through the Military Administrator. The meeting will give you an idea what the wretched fellow thinks of me. It will help me assess what they're up to.

There's a flash point in Ogoni right now. The gas pipeline. The Govt refused to give visas to our own consultants who were to help us review the EIA [Environmental Impact Assessment] which we had insisted on, and which is required under Nigerian laws. Govt. had already paid for the study as we demanded, and I did everything to convince the Govt. to issue the visas. I spoke to the Nigerian High Commissioner and to the Minister of Petroleum. And then I was arrested.

What we have to do is to stop Saipem[2] (the Italian Contractors) from walking over Ogoni dead bodies for their profit. I've written to Mr. Ohlsen[3] and we are going to lobby all EU Embassies in Lagos & the EU itself.

I was told to apply for some relief funds from the EU (regardless of whichever else we had previously applied for). Alfred Ilenre[4] brought the news, but I don't know whether this is correct. We have applied anyway. You may check this out.

All the best, and let me hear from you soon. I think you might be able to see me if you try hard enough.

Regards, Ken.

1. McCarron had returned to Ireland on 16/8/94 and planned to stay there in the longer term to work with issues related to the Northern Ireland conflict. While visiting the Trócaire office, she was asked to return to Nigeria to facilitate a review of the application to the EU for humanitarian aid for Ogoniland. On her brief visit there, she was prohibited from visiting Saro-Wiwa in detention but they continued to communicate by letter.
2. An oil and gas industry contractor.
3. Gerald Ohlsen, Acting High Commissioner for Canada in Nigeria.
4. General Secretary of the Ethnic Minority Rights Organization of Africa (EMIROAF).

October 19, 1994

19/10/1994

Dear Sr. Majella,

I got your diary which ended on 13th October. And considering that up to yesterday, there was no news of you having got to PH, I'm hopeful that this will get to you before you return to Europe.

Thanks a lot for your letter and for all the information which you gave me. It was most cheering indeed, and brought home to me the loss which I feel at your being so far away. I wonder if you have checked with [initials] if all that you sent me did get to Lagos. Because I don't appear to have received anything from you since you left Lagos. And I wonder if you did receive my own letters. You do not appear to have indicated so in your recent communication.

There isn't much to write since I think I wrote you less than a week ago. I hope to hear from you before you return to Ireland. I haven't heard from my son or daughters and I'm worried about that. Particularly about the girls' A-level results.

I'm in good spirits. I hope that the Court of Appeal tomorrow at least orders these goons here to allow me visits by family, doctor and lawyer. I don't expect anything better until the Govt are able or are unable to build the gas pipeline for which they are holding me.

Cheers, Sr. and God bless you.

Ken

P.S. I did receive a message after my devotional hour on Tuesday 11th October of coming good news. I hardly thought it was going to be the Right Livelihood Award[1] which has cheered me a great deal. The message I got spoke of a wider role for me in which my MOSOP and Ogoni experiences would come

---

1. The Right Livelihood Foundation is a Swedish-based charity which promotes education, research and practical activities that contribute to global peace and justice. The annual award offered by the Foundation has become known as "the Alternative Nobel Prize". It was established in 1980 following a failed attempt by journalist and philatelist Jakob von Uexkull to have the Nobel Foundation create new prizes in the areas of ecology and poverty elimination. For more information, see <http://www.rightlivelihood.org/>.

in handy. The Award has raised the morale of the Ogoni people and is a psychological defeat for our enemies.

By the way, you might take away the recent NTA [National Television Authority] Port Harcourt propaganda film done to discredit me. The most important bit is the point where the Commander of the Task Force [Paul Okuntimo] describes how he tortures Ogoni people. Europeans should see that clip.

October 24, 1994

24/10/1994

Dear Sr. Majella,

A million thanks indeed for yours of 14th Oct-21st. It was, as usual, most informative and made me miss you more than ever. You are the only one who gives me an almost complete picture. Had you gone to Ogoni, I'd have got a complete picture—local, national and international. It can be terrible here when one cannot phone or discuss with others. I spend a lot of money keeping the lines of communication open because I need every bit of information on time. As the Ogoni people learn to fight from the underground, not many of them realize the value of detailed information to those of us in detention. And we do not have a culture of writing. But because I reply to letters as soon as I get them, they are now learning to write to let me know what is happening. No one in London has written me—even my son did not write after his first letter—and my cousin who heads MOSOP in the U.S. has not written either. I only hear from him on the pages of newspapers.

The RLA prize [Right Livelihood Award] was most welcome. It encouraged the Ogoni people a great deal, legitimized MOSOP as a non-violent, and environmental and human rights organization, and the prize money will ease things a great deal for me. I don't see Shell and the government allowing me to travel—they must dread what bombs my presence will drop in Europe as I'm supposed to address the Swedish Parliament, the European Parliament in Strasbourg and another meeting in London. There or not, my words will ring through all the places. If I can't make it, I intend to ask my son to represent me. But somehow, I'm hopeful that I'll be there. If I'm not, then it is in Ogoni interest that I should not be. God's will.

As the days go by, I get the more convinced that the cause will win. I remember your encouraging me in the early days of our meeting, saying how because I had a certain independence of means, I might well be the only activist capable of giving Shell a run for their money. When I think how far we've gone on very thin resources, I have cause to be grateful to God. And no matter what Shell does or says, they've been in rough waters since July 1992 when I advised the Working Group on Indigenous Peoples[1] in Geneva. I am grateful to all those of you who have rallied round the Ogoni cause—UNPO, Greenpeace, International Pen, etc. And there must be better news on the way. I should mention the Bodyshop, of course. You probably know that they nominated me and MOSOP for the RLA Award. I have sent an appeal to President Carter asking him to intervene and resolve the conflict. Someone of his reputation would make quite a difference. My cousin in America has been quoted as saying the MOSOP (USA) would sue Shell. Exxon had to pay 5 billion USD for the oil spill from one tanker in Alaska. By the time we've created sufficient awareness internationally, it should be possible for us to find assistance should we wish to sue.

Sr., I don't know if my wish is father to the next thought. But somehow, there is hardly a thing which I have undertaken which has not been successful. I see the hand of God in it all, and am grateful to Him all the time. My preparation for this struggle was a long one and it was made simple by the fact that I did not know that the Divine Hand was preparing me all the time. It is only now when I look back and begin to put the pieces together that I can see the progression from Administrator for Bonny, Commissioner in Rivers State, businessman (of some success), television producer, publisher, writer to activist as preparation for a task that would have been daunting if I had, for one moment, stopped to think or analyse its implications. And believe you me, I did not ever plan any of these things. They just happened to me. Only the decision to be a businessman after I had been sacked from the Rivers State Cabinet in March of 1973 was, consciously taken—but I didn't have much of an alternative![2]

Yes, God has been very good to me. I've lived a charmed life. Consider that my 90 year old father and 73 year old mother go to court each day my case comes up; that you suddenly turned up to do the work among the diplomats & others in Lagos which we could not have done; that the Ogoni people have been so entirely supportive and have borne the pain of the struggle

---

1. The author is referring to the UN Working Group on Indigenous Populations.
2. Saro-Wiwa was Commissioner for Education and later Commissioner for Information and Home Affairs in the Rivers State Cabinet between 1968 and 1973. After the Civil War (1967-70), he became increasingly vocal in opposing the appropriation and misuse of the state's petrodollars by the nation's military elite under General Yakubu Gowon. This resulted in his expulsion from government in 1973.

without complaint, that my brother Owens is right there always—he even had the wisdom of leaving Port Harcourt in time; that I have family members like Barika and Komene[3] (whom I trained through secondary school, by the way) and other relatives who are extremely loyal and hard working, and you will agree with me that the finger of God is in it. And the struggle was really quite inevitable. The Ogoni would have perished otherwise. Even if they were to perish now (and it's not possible), it would not have been without a fight! Oh, how proud the Ogoni people now feel! And not without justification. The RLA Award, I'm afraid, has even won them the envy of their neighbours and other Nigerians. A bad thing, but we'll get over it in due course.

We need your prayers, and your endeavours and I keep praying that you remain in good health to work for us! And find some other fulfilment. God bless you.

I am ashamed to say that I do not know the full details of the 3 projects being funded by the EU. I did see, by chance, a fax from Trócaire to the Bishop and that's how I knew about ECHO [European Community Humanitarian Aid Office]. What are the others? Please do not worry that you are being shunted aside after all that you have done. God uses us in his way and it may well be that your assignment was to bring the projects about. I dare hope that many more projects and other assistance will be forthcoming. I hope that we can work for these other ones. After all, the Ogoni need a million projects. That is why I have not been worried about who is handling it or not. The important thing is that help should get down. You did warn me, anyway, about the politics of aid. What I wonder is, does the Diocese have the expertise to handle it all? My brother and Ben probably do not realize that MOSOP was not to be touched with a barge pole. Legitimizing the organization, given the enmity of Shell and the Nigerian Government was always going to be a problem. We do not even have a bank account! And they have to realize that we will have to fight for a long time. Why worry about these early pains? We will now begin to assemble the atrocities as required by Amnesty International. But to what purpose? Just to keep the statistics?

---

3. A member of Saro-Wiwa's extended family who was in Rome undertaking a course on research and documentation when Saro-Wiwa's arrest took place. He did not return to Nigeria. McCarron helped him to acquire a scholarship to undertake an MA programme in Dublin jointly at University College Dublin and Kimmage Centre for Development Studies.

I will write the Ecumenical Committee for Corporate Responsibility as you advised—through you. I have some ideas which they may find useful.

Did the court grant Shell the injunction they have requested on *The Drilling Fields*? I'd like to know.

As far as I am concerned, Shell should lose its mining lease in Ogoni. They may be pretending that they do not want to return to Ogoni. The fact is that they have 500 million barrels of oil on secondary drilling at K. Dere;[4] they only last year awarded a 550 million USD contract to some organization to design the gas collection throughout Ogoni and the K. Dere field was to help in the natural gas plant at Bonny. No, Shell are merely hoping that the government will succeed in "pacifying" the Ogoni and then they will move in proudly and calmly to continue to steal. They are in for a fight they will never forget. Luckily, I'm no longer alone. Several Ogoni youth are now learning the ropes, and if only they could get further exposure, they would be able to continue the struggle even in my absence.

I'm happy that the question is being raised in the Irish parliament.[5] The more air it receives, the better.

I did hear the BBC broadcast. The lady travelled in the delta with Oronto Douglas[6] who helped Nick Ashton-Jones prepare his report. Oronto is a lawyer and committed to the Niger Delta—his home is one of the six places studied.

I'm sorry that I did not get to see you when you came to Port Harcourt.

4. 132 A village in the Gokana district, Ogoniland.
5. The September attacks on Ogoni people and their property by members of the Nigerian armed forces were raised in the Irish parliament [Dáil Éireann] on 18/10/94. The then Tánaiste [Deputy Prime Minister] and Minister for Foreign Affairs, Dick Spring stated that he had supported a proposal, made by EU representatives in Lagos, for a fact-finding mission to be sent to Ogoniland to assess the situation. The crisis in Ogoni was discussed on numerous occasions from then until 1998, when it appears more intermittently in the parliamentary records. For more details, see <http://www.oireachtas-debates.gov.ie/>.
6. A leading Nigerian attorney and human rights lawyer who was among the team of lawyers representing Saro-Wiwa during his trial. He is co-founder of Environmental Rights Action and of Friends of the Earth Nigeria, co-author with Nick Ashton-Jones of *The Human Eco-Systems of the Niger Delta* (1998), a handbook of Environmental Rights Action (ERA) and co-author with Ike Okonta of *Where Vultures Feast: Shell, Human Rights and Oil in the Niger Delta* (New York: Sierra Club Books, 2001).

I had hoped I might, even though I know how impossible it was. But we'll certainly meet again, and I hope it's soon.

I'm not going into partisan politics. What I meant is that I would be taking a wider role in the nation's affairs–expanding the Ogoni struggle to other parts of the delta and beyond. I could never be a part of whatever Abacha is planning for the future. What I want to see, and what I will always argue for is ERECTISM – ethnic autonomy, resource and environmental control. If this comes to pass, then Ogoni will be free and it is to them that I wish to dedicate the rest of my life. And I hope that that can be an example to other ethnic groups. The translation of my dreams into reality. Nothing to do with partisan politics.

I don't know if this letter will get to you before you leave for Dublin. Whatever the case, take good care of yourself and may God bless you.

Ken.

P.S. The programme for Sweden is as follows:

December 5. London. A meeting possibly in the Houses of Parliament.

December 6. Travel to Stockholm. Informal Reception for Recipients and RLA representatives.

December 7. Press Conference.

December 9. Award Ceremony, Swedish Parliament.

December 10. Seminar?

Thereafter, possibility of meeting European Parliament in Strasbourg.

Video presentation of recipients to be organised by Robin Sharp [London address provided]–Film materials, video, photos or posters to the above-named.

Terry Ndee[7] might be requested to contact Greenpeace for photos and I wonder if *The Drilling Fields* will be useful.

---

7. An Ogoni member of MOSOP who was based in London.

October 27, 1994

27/10/94

Dear Sr. Majella,

Thanks for your diary of 22-26/10. Don't be surprised at the paper I'm using. There has been a crisis here. I "lost", on Tuesday evening, all my writing paper, books, photos etc. and, unfortunately, two or three letters to Major Okuntimo, Commander of the Ogoni Murder Squad. I was the victim in a struggle between two Army Captains–the departing one (who had not wanted to leave) and an in-coming one (who badly wanted to be here). It's a long story, but in the end, the in-coming one used the "lapses" of the outgoing jailer in allowing me certain things to outwit his former classmate. I rescued my pen in the last minute–my radio too. But Nick's beautiful report went and much else. Fortunately, I always return letters within 24 hours so that it was only those sent on Monday and one other from Mon's father-in-law that were seized.[1] I just hope those ones are not now in trouble. I've got to remind Appolos [Dr Appolos Bulo][2] not to receive letters which bear addresses. We all have to learn how to work in the "underground".

Thanks for coming back. And please do not be in a hurry to return to Lagos. We need you. I had not gotten down to writing anything for the [RLA] Award because I'd not known precisely what was to be expected. When I finally got it, I'm now under stress with the events of Tuesday 25/10. I now need a copy of the Award letter (I wonder if you or the office have it), and you have to stay around to read and correct what I write–the view from the Prison can be quite jaundiced and illiterate, and I have no reference books from which I could quote–before you take them away.[3]

It may be possible to keep the lines of communication open–if Appolos can find the money to pay the new jailer. I was just about to give the man a bit of the stuff I'm made of and refuse to accept his attempt to extort money from me. But the Award Statement might make me mellow–how can one accept

---

1. Owens Wiwa is also known by the familial name Mon probably because he was born on a Monday.
2. An Ibo assistant to MOSOP working in the Port Harcourt office.
3. McCarron brought Saro-Wiwa's acceptance speech and other items to Stockholm for the RLA Award ceremony.

evil? Or co-operate with it? Or am I just getting it out of the way–"stepping it aside"? One is so very vulnerable here!

Don't worry about the Ogoni jujus. That bit was meant to frighten the Nigerian soldiers. I understand their mentality. And quite a number of other Nigerians–including the Ogoni people. I really had a good chuckle when I inserted that bit as an afterthought. The Born-Again Xians [Christians] have been worried–I read something like that in a fun-column of the *Sunray* newspaper. But it's O.K. Remember Hamlet. "There's more in Heaven and Earth, than is compounded in your philosophy, Horatio". I'm using every non-violent weapon to face very violent people. Break your ribs.[4]

Read some of the letters I receive here–which I always return to the office–they will give you an idea how the faithful feel. Appolos has them all.

I'm asking President Jimmy Carter to intervene in the Ogoni situation.[5] I'll be keeping it in the quiet for now and the Lagos office should be advised not to talk about it to anyone. You may look at the letter–for your information. I cannot make any emendations now so don't worry if it's not perfect.

Don't expect anything from the court. This matter is political, and the military do not care for the judicial system. The only thing that can save the Ogoni people is the conscience of the West. If the embassies in Lagos pressure Abacha, yes, anything can happen. I fired off a challenging letter to the National Security Advisor (Asmaita Guarzo) weeks ago. Ask Appolos for a copy. I understand Komo is doing a tour of the local government areas. I hope it is a "goodbye" tour. He has really goofed and presented me with a bonus. I intend to call, in my Award Speech, for an International Rescue Mission to save the Ogoni people and environment.

Just remember this. The embassies in Lagos are very important. No EU

---

4. A Nigerian equivalent of the expression "split your sides [laughing]".

5. Jimmy Carter and other international figures had intervened in Nigeria in March 1994 when the Army Chief of Staff, Olusegun Obasanjo, was arrested and threatened with execution for involvement in a failed coup. An outspoken critic of military rule, Obasanjo had been the first Nigerian military ruler to hand over power to an elected civilian government in 1980. At Carter's behest, he was released but imprisoned again several months later that year. He remained in detention until Abacha's death in 1998. Saro-Wiwa understood the power of international pressure to protect critics of the dictatorship and hoped to mobilize international support to protect himself and other MOSOP members. McCarron does not know whether or not he received a reply from Carter.

project can succeed if the Nigerian Army are still around. They will loot the place. Their soldiers are hungry! What do you think a Nigerian soldier earns?—800.00 naira a month. And that's after 27 years in the force. I've had to feed the soldiers who guard me! The officers are even more ravenous and thieving. They're all armed robbers. So the thing is to get them out of the area so that some peace can reign. You only have to remember that all we've had has been state-violence. The Ogoni people have been quite calm. I even suspect that Kobani and others were murdered by the security agencies in order to justify some of the reports that had been submitted by the security people in support of the Constitutional Conference. We (Ledum Mitee[6] and I) have met soldiers who are prepared, if they have the protection, to talk about what instructions they had, who looted what, who killed whom. The EU & the Americans must strive to get the Nigerian soldiers out!

Did Barika's letter to the EU Delegate go to other embassies?[7] It should. (He spelt "psychopath" wrong). I'll wait to hear from you.

Ken.

6. A MOSOP leader who was arrested with Saro-Wiwa. He was ultimately spared execution due, in part, to the pressure of international lobbies.

7. Barika Idamkue had been lobbying Western embassies to pressure Abacha to release the MOSOP activists.

October 29, 1994 (1)

29/10/94

Sr. M,[1]

Thanks a lot for the corrections to the Speech.[2] I'm also sending the 5-minute address. The third one—the 30-minute paper should be ready by the end of today. I must appeal to you to wait and take them all with you. Please!

I may still need to correct them before the 14th of November because I will ask Professor Claude Ake to look all three over. I rely on his judgement a lot. If it be the case that some changes are required, they will be made, and you will be duly informed—before 14th November.

You are very kind in the way you commiserate with me. But you must know that I'm quite tough. Writing a speech of this nature is normally a joy to me and I was not exhausted after writing it. I was, on the contrary, quite elated.

The fact that the physical conditions here are quite good, that I know there's a host of Majellas & the ordinary Ogoni people caring, praying for me, my belief in my mission and God's purpose for me in that mission, these have buoyed me considerably. My moments of depression have had more to do with the political situation in the country, worries over the Ogoni and such-like, than the fact of my confinement. I miss my family, of course, but as I say, it is a fitting price to pay for the joy of others.

This may not sound very nice, and most people around me do not want to hear it (so don't tell Hauwa), but I have assumed for quite some time that death cannot be very far away from me. The period of confinement has helped me to "a closer walk with God". I was always religious, but the spiritual side of me was not properly developed. I am developing it now. Somehow, I do feel that God is in charge and that all I need to do is work hard, take advice and train & encourage the younger Ogoni to carry on from wher-

---

1. A note added to the top of this letter states: "I've made photocopies of the 'To Whom It May Concern' available to the Lagos office for transmission to the Johnstons etc." It referred to letters publicizing the Ogoni cause which McCarron had agreed to send. The Johnstons were a Scottish couple who had lived in Nigeria for many years and were sympathetic to the MOSOP cause.
2. Saro-Wiwa's Acceptance Speech for the RLA award ceremony in Stockholm on 9th December 1994.

ever I may stop. Sr., I'm at peace with myself. My guards are on my side, and the Asst. General Manager of the NEPA station (an Igbo of the name Godwin Oboli) has been wonderful. Replying to the letters I get keeps me busy and those letters give me pleasure too.

Yes, I have everything to be thankful for, and do not forget that I've been here only 23 weeks now. Mandela & Walter Sisulu were there for 26/27 years. How can I complain?

I'm going to ask to be allowed see you so we can pray together on Sunday, even if it's for ten minutes or so. Keep smiling.

Ken.

October 29, 1994 (2)

29/10/94

Sr. M,

1. I almost forgot this! For the London meeting of the RLA as demanded, it would be well worth it contacting Glen Ellis of Oxford who made *The Drilling Fields.*[1] He has a load of material which he has not used, including me reading from A *Forest of Flowers.* Kweku also has a film of Ogoni women of FOWA [Federation of Ogoni Women's Association] at work. Appolos has the address and telephone number of Glen Ellis.

2. [2]The flag and poster of MOSOP are also available in the office and you should please take a number of each.

3. MOSOP singlets are also available—there are probably some in my house. Elfreda will assist you get them.[3]

4. The Rivers State Govt's "MOSOP" film has a section where the Commander of the Internal Security Task Force [Col. Duada Musa Komo] describes his methods. I asked that a clip of this be made. You should please demand it. There's also a place where he accuses me (falsely, of course) of misusing funds donated from abroad. That clip may be useful.

5. Greenpeace Communications Ltd. have those photographs of the devastation of Ogoni.[4] I have asked for colour reproductions of the ones we have in black and white. You may renew the demand for them. I will be prepared to pay for them. I heard on radio this morning that they are cash-strapped, and have had to lay off some staff. They owe me £700.00 which they've refused to pay since January 1993. But that does not mat-

---

1. McCarron had agreed to meet representatives of the Right Livelihood Foundation in London to hand over Saro-Wiwa's Acceptance Speech and other materials for the award ceremony in Stockholm on 9th December 1994.
2. Enumeration in the original letter starts at 2. There is an indistinct number 1 alongside the first paragraph which appears to have been added as the list of items grew.
3. Elfreda Jumbo, a friend of Saro-Wiwa in Port Harcourt.
4. Copies of these photographs are held in the Ken Saro-Wiwa Collection at the National University of Ireland Maynooth.

ter. The RLA Foundation may be willing to pay, I don't know. In any case, I'm desirous of having these photographs for the exhibition in Sweden. I'll send a cheque to cover costs. The Award money can be used for it.

6. For the same exhibition, we need a single copy of each of my books (25 in all published as of today–7 are in MS [manuscript] stage). Between PH [Port Harcourt] and Lagos, they should all be available. Whatever is not available (*Transistor Radio* may not be), ask Deebii[5] to collect from Heinemann (Lagos Branch). Collect the pamphlets too.

7. There is a wretched film of me in the Civil War days done by a German Company. A "poor" copy is in my house here. Ask Elfreda for it. Some clips may be found useful.

8. It is possible that the Swedish Embassy will willingly put all we need for the exhibition in a diplomatic bag and send it all to Sweden. The Americans did it for me the last time I went to the United States at their invitation. This could save us all the embarrassment & weight carrying these things around.

9. In case I'm not able to attend the ceremony, a team of 2 will be leaving Nigeria for it. But I'll ask my son to read the Acceptance Speech and collect the prize on my behalf while those coming from Nigeria will be on hand to attend press conferences and do whatever else.[6] The last I heard of the young man he was in the U.S. He may not have sent letters to me because I had said I would let him know the best way of doing so. If you find him, please introduce him to the [initials] route. It will help us all a great deal. I'm sending him a letter as to how he'll get to Sweden (money for travel, hotel etc.)

10. I will also be asking my cousin Vincent[7] to see if he can make the journey to Sweden. The London Ogoni as well. It would be wonderful if we could get a small Ogoni crowd in Stockholm. There's only one Ogoni man in Sweden. I expect he'll be there to assist.

---

5. Deebii Nwiado managed Saros Publications in Lagos.
6. Saro-Wiwa ultimately decided to assign the task of reading his speech to a member of MOSOP. McCarron believes that he changed his mind because he wished to demonstrate the relevance of the award to the Ogoni people rather than to just himself or his family. Simeon Kpoturu of MOSOP read the acceptance speech in his name in Stockholm on 9th December 1994.
7. Vincent Ideymore, President of MOSOP in the U.S.A.

11. I hope that you can make it. Speak to the Bodyshop. They may be willing to provide a ticket & hotel accommodation

12. This is a load of work, but I expect you, as usual, to do it all and even better. I'm sure you have other ideas that will enrich the Stockholm exhibition on which I lay great store. Glen Ellis, if he's in Oxford (pray God he is) will be of great assistance because he also has some good photographs. (By the way, some of his photographs are in my briefcase—large ones—ask Elfreda—I saw them the other day when I had the briefcase brought here for a moment.) You have seen them in Lagos.

13. I don't know if some of the magazine covers dealing with the Ogoni situation will be useful. A number of them are available as well as pen-portraits of my happy self—both local and overseas. They are there in the file.

14. We certainly will need a map showing the relative position of Ogoni in Nigeria and Africa. The newspapers in the U.K. often provide that quite easily, if we cannot get it locally.

15. In case Glen Ellis is not available, I wonder if Robin Sharp [of] the RLAF[8] Research Section in London will be able to use *The Drilling Fields*?

16. My mind is dashing about the place, so if it's not orderly, don't worry. I'm taking a chance to write to General Oladipo Diya—Chief of General Staff—to ask him to get me out of here and enable me to go to the Award Ceremony.

Regards,
Ken.

---

8. Probably a mixed acronym: Right Livelihood Award Foundation.

October 30, 1994

30/10/94

Sr. M,

You did ask if you could "contact" my London family? The answer is "Yes". I think they will value it.

I am sending you a letter for Robin Sharp, formally introducing you to him as acting on my behalf. You did indicate that you might find it necessary.

I envy you the meeting with our people. They must have been pleased with the song of praise in *Tell* magazine listing the Ogoni as the most politically aware in Nigeria. The article was good, although it had many grammatical errors. The *Newswatch* report was probably more "in-depth"–as regards my person.

The big essay [for the RLA Award Ceremony] is sent here in long hand. I hope that you can wait to have it word-processed. It need not come back to me. You can correct it and take it along. Please watch out for typos. Emeka tends to make mistakes on crucial words. In the Acceptance Speech, "Ogoni Woman" was typed "Ogoni Women" which altered the meaning radically. "Wasting Storms" came off as "Washing Storms" an entirely different thing.

I had expected Appolos here last night when I would tidy up plans for meeting with you with my jailers. He failed to turn up. I hope he has given you the long letter I wrote yesterday asking you to please do a million things for me.

Have a good, safe journey back to Dublin and may God bless you and keep you. I hope to see you soon. Did you see my poem, "A Walk in the Prison Yard"? I've asked Deebii to give you £20.00 for stamps for any postages you have to make on my behalf in U.K.[1]

Ken.

P.S. The cheque for the photographs from Greenpeace is included herewith. Also a cheque for my daughters. They are to split the money equally. I need a new cheque book. There may be one in the house but a request note is

---

1. This sentence was added in note form alongside the signature.

included herewith. There is need for them to look through my mail, pay cheques which may have arrived into my a/c. There is, I believe a cheque for £2000+ meant for the a/c of Saros International from my book distributors. All book orders to be re-directed to African Books Collective in Oxford. Ken Junior should be able to attend to all these if he is in London. The insurance for the house may have run out. It should be renewed–the bank should be requested to pay all premiums. The same could apply to the car–possibly.

22 November, 1994

22/11/94

Dear Sr. M,

Greetings. I hope you did get back safely with all that load of work I inflicted on you. You can see how much needed you were in Nigeria! You are truly a God-send. I had hoped that you would leave behind a few thoughts for me before your departure, but you must have been very busy.

In the continuing Ogoni Saga, you must have heard of the setting up of a military tribunal to try us—under a brand new Decree which we have not yet been able to see. The 3-man Tribunal has been named and were to have been sworn in yesterday (Monday 21/11). It's really a hangman's work. Our only hope is that the international community may be able, diplomatically, to frighten Abacha out of any mischief as they did with the Treason and Treasonable Felony Decree 1993 which was aimed specifically at me and the Ogoni people. This is terribly sickening. Shell are finally succeeding, it would appear, in shutting me up. But that will surely not be the end of the matter.

I understand Glen Ellis is in Oxford at this time. I hope that you have been able to make contact with him in regard of the Right Livelihood Award function. Barika [Idamkue] and Batom Mitee will be going for us, since I and Ledum Mitee cannot get there. I expect that my cousin, Dr Vincent Idemyor, who's head of MOSOP in the U.S.A., will be there with his Sr. [sister], Bridget and that some of the people in London will be able to go too, along with my children (if they desire to travel). One of the activists (probably Godwin Poi from London) should read the Acceptance Speech—and not my son as I had earlier indicated.[1]

I hope that you have been able to put the things for the exhibition together and that all will go well. I wish you the very best and in case I do not get to you before Xmas, wish you a Merry Xmas, & a Happy New Year.

Sincerely,

Ken.

1. See n. 59 with the letter dated 29/10/94

December 24, 1994

24/12/94

Dear Sis M,

Thanks a million for your diaries of August, November and early December. It was really wonderful to hear all about the Stockholm Award Ceremony from your careful, observant pen. Before I forget, I do not know who G is, and the connection with Channel 4. What's he supposed to be doing? Could he be Godwin Poi (formerly OCA/UK [Ogoni Community Association/United Kingdom] Resident?) And what films is Glen now producing? I expect there's the one for Canada, but for whom is the other one?

I may have missed it, because I have had my radio seized for some time now, but the RLA ceremony was not reported by the BBC African Service. I did hear Barika on the World Service *Outlook* programme but I had expected the African Service to talk about it.

I think that we've done pretty well with the NGOs. The thing now is to get the support of some governments. My brother tells me the Swedish Govt is pliable and so seems the Irish Govt. I believe that effort should now be concentrated on this. Govts. can raise the matter at the UN and at other fora. They can also assist us set up offices and provide funding. The Swedes & the Swiss are quite good at that sort of thing. You should also find time to visit Rudy Drummond who sits on the Nigerian desk and the Foreign & Commonwealth Office in London. These governments can also encourage the media to support us–as they can discourage them. The British have been cool towards the Ogoni cause (and supportive of various military dictatorships in Nigeria). Junior Saro-Wiwa appears to have reached Prince Charles (through Olivia[1]) and since the monarchy has shares in Shell, his intervention

---

1. Olivia Burnett, English girlfriend of Saro-Wiwa's eldest son, Ken Wiwa. They married in 1996.

could help. He (Prince Charles) was in school with William Boyd[2] and the latter might be prevailed upon to drop a note to the former?

I'm happy that you've met up with the family at Epsom. Maria [Saro-Wiwa's estranged wife] was always a good "Catholic" girl and may easily have taken to Holy Orders if she had had the encouragement. I could have been a better husband to her if my life did not get caught up in so many wars. I was always very proud of her and anxious for her happiness. She has passed through a lot that is not pleasant (deaths, illnesses etc.), and what comfort you can give her will be very much appreciated by me. I'm happy that you are finding Zina pleasant. I wish her Sr. [her sister Noo] were not so withdrawn but I'm happy that she's now at King's College.[...]Yes, Tedum was a gem. The memory of him hurts me deep, real deep. May his soul rest in peace.

We've decided to challenge the setting up of the Tribunal in court. We may, we will lose the case, but it will serve to delay the kangaroo trial and win us the sympathy we need. Govt are aware of this and are already talking of a secret trial. Whatever the case, locking me away for some time will not solve the Ogoni problem. I'm sure that we will win in due course and that my ideas on the structure of Nigeria will supervene.

Your letter did not worry about your health, so I must presume that everything is alright with you, and rejoice with you on that score. I'm happy that you are able to give so much time to the Ogoni cause.

I'm surprised that any Ogoni person should be worrying about where funds [RLA prize money] are to be paid to. None of these can take care of a huge sum of money. The only one who can use it properly for the proper cause is myself and I will definitely oppose anyone else, or any group tampering with whatever money comes to MOSOP. I think they all ought to know that I've funded and will continue to fund the struggle from my resources. And I'm the only one who can produce a properly audited account that will satisfy everyone. So if the matter ever arises, you should ask them to rely on my sense of probity, my reputation, and the fact that I have enough money of my own not to fool around with a common purse.

2. British novelist and screen-writer, author of six novels, including A *Good Man in Africa* (1981), as well as numerous short stories and screenplays. He was born in Accra, Ghana in 1952 and grew up there and in Nigeria. He corresponded with Saro-Wiwa while the latter was in detention and composed the Introduction to the Longman 1994 reprint of *Sozaboy*.

We may be requiring your assistance to get some U.S. dollars into Nigeria from my account in London. IF this need arises, my brother will contact you and please let him know if you can do it for us.[3] The fact is that we need to get proper naira value for any hard currency we have. We may have to bend the laws; after all, the laws are being bent to drive us to extinction.

Best wishes for the New Year.

From me,

Ken.

3. McCarron notes that this need did not arise

January 15, 1995

15/1/95

Dear Sr. M,

Thanks a lot for your diary from 6th Dec to 10th January. It was, as usual, a treat, a mine of information. Somehow, I never got the attachments. I thought I'd get the *Guardian* article, the Wole Soyinka piece and anything else of importance but I was not so lucky.

I'm really happy that you are finding so much help for the Ogoni people. Somehow, it's exhilarating to find a specific assignment for oneself– something that can last a lifetime, which is different to what everyone else does or can do, and whose success or failure can touch a large number of people. I don't know if I make myself clear. It's one thing campaigning for the environment, it's another campaigning for the environment rights of the Ogoni people. This is what gives me the kicks.

Specificity. And it might have been better for me if I were not Ogoni. I should feel like a missionary. Now I think I'm doing what I ought to do as the son of my father. A duty. But I'm happy that I got round to it in the end.

Sorry about the dearth of news from this end. The newspapers are all under tremendous pressure and radio & television being govt. controlled are useless to us. There's not much happening anyway. Just the old stone-age dictatorship strutting about the stage like a blasted peacock in dark goggles. Such a walking insult to the Nigerian! I feel so ashamed to be Nigerian.

Ogoni Day was a success here at home in spite of the blandishments of the military. FOWA [Federation of Ogoni Women's Association] played a very impressive role. 15 of them ended up in detention in the military camp at Kpor.[1] There were processions and dances in individual villages as I had ordered. In Bori, the vulture who acts as chairman of the Local Govt. Council who ordered the women arrested was driven by the women from church when he went the following Sunday to read the first lesson on Army Remembrance Day. The resilience of Ogoni women is admirable. My mother continues to host, each week, meetings of Ogoni women from fourteen surrounding villages. I've seen her twice since I got to Bori Camp here. She lost one of her 2 Sr.s [two sisters] a month ago or so. But she's bearing up well. My

---

1. A village in Ogoniland.

father had a surgical operation recently. But I'm told he's now okay. There's a video of Ogoni Day & we'll be sending it on once it has been edited. I sent a speech which has been published locally in part. It was read to applause at one of the centres.

Tomorrow, we go to the Tribunal. I know it's a kangaroo court and I know they'll be shutting me away for a time if not forever. But I'm not worried about that. I believe that it will give point to my cause and give it the world-wide publicity which we so badly need. Troops have been sent to Ogoni to stop the people from turning up in large numbers at the trial which is to be held in Government House. Government House! That is like holding the trial in the bedroom of the hangman. My younger Sr. [sister] who's a lawyer in Zaria has come down and I expect my parents to be there, as usual. I've seen a handout asking all people in the oil-producing areas to protest the trial. I asked Gani [Fawehinmi] (who's our lawyer) to go to the Lagos High Court to Challenge the Tribunal. He finally did, rather late. So the trial will go on, but he can argue his suit on the 26th in Lagos. A pity we could not delay the start of the Trial as I had wanted to do.

We've got an incriminating document belonging to the Rivers State Govt which I'm asking to be sent to you. I hope it gets published and I'd like to know what is the public reaction to it. UNPO appears hesitant to use it, doubting its authenticity. But I think it's authentic, as Govt. has followed the plan absolutely. So, we'll be at the Tribunal singing the MOSOP marching song.

B [probably Barika Idamkue] brought me a copy of Mandela's auto-biography *Long Walk to Freedom* and I really enjoyed it over the New Year Holidays. I felt very energized by it. It's so fictional in quality. And it's very well written. Short chapters in various sections. You must get a copy.

I'm going to try to get the Tribunal to move me either to prison or to hos-pital. The Nigerian Army is brutal, greedy for power and money. I'd like to be rid of them forever. If I don't get that, I'll be really mad. I need to get to a place where I can write or at least use my computer.

Your diary did not say a word of my daughters. I hope they remain close to you. I miss them a lot. Also my daughter, Adele, who's with a friend of the late mother at Derby. She's really sweet, younger than Singto who was here over the Xmas. She came to see me in prison and was I glad to see her! I hadn't seen her for almost eight months. She had grown much taller and slimmer. And she's as enthusiastic as ever. Hauwa was also here throughout Decem-ber with young Kwame (who looks so very much like me). I was able to hold him in my arms twice or thrice. I was not important to him, though. I'll write again as the trial progresses. I'll be having quite some fun in court. You have my love & admiration.

Ken.

February 7, 1995

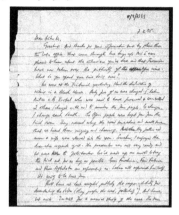

Maynooth University Ken
Saro-Wiwa Archive.

7/02/95

Dear Sr. M,

Greetings. And thanks for your information sent by phone through the Lagos office. That came through two days ago. And I was pleased to hear about the interviews you've done and that Trócaire have taken over the publicity of the Ogoni issue. What do you spend your time doing now?

We were at the Tribunal yesterday. Had the distinction of riding in a Black Maria. Only five of us were charged. I, Ledum Mitee & B. Kiobel[1] who were said to have procured & counselled 2 others (charged with us) to murder the four people. 4 charges, 1 charge each death. The Ogoni people were kept far from the trial room. They massed along the road somewhere and made sure that we heard them singing and cheering. My father and mum & wife were allowed into the room. Somehow, I enjoyed the dramatic aspects of it. The prosecutor was not very ready and we were to filibuster.

We've made up our minds to drag the trial out for as long as possible. Gani

---

1. Barinem Kiobel, one of the "Ogoni nine" who were executed by the Nigerian military government on 10th November 1995.

Fawehinmi, Femi Falana and Olisa Agbakoba are representing us. Ledum will represent himself. It's going to be some fun.

Shell have at last accepted publicly the responsibility for devastating the delta (they prefer the word, "polluting"). And have set aside 2m USD for a massive study of the area. The sum is too small, & is to be administered by a Steering Committee headed by Gamaliel Onosode—a man of honour, Chairman of both Dunlop & Cadbury of Nigeria. MOSOP will be responding to all that. In their press release, Brian Anderson [Managing Director of Shell Petroleum Development Company][2] says, "It is important to have a comprehensive and independent survey on Nigeria's delta region. Recent politicised and emotive campaigning has clouded some very important issues concerning the development of this region". I consider this development a victory for the Ogoni people & MOSOP.

You may have heard of my having won the 1995 Goldman Prize awarded by the Goldman Environmental Foundation of California and worth 60,000.00 USD. Quite a welcome development. It will be officially announced on April 17.

An aspect of the trial is, we're trying to obtain bail or at least to move from Bori Camp to the wretched PH [Port Harcourt] prison so at least to be away from the control of the military. That comes up on 21st February.

I hope that you can send or fax me any articles which appear in the dailies out there. I understand that there have been 2 things in both the *Observer* & the *Guardian* of London. The radio waves were full of the trial on the 6th of February. The BBC World Service made it the first item in their news bulletins from 1pm till 10pm or so and there was an interview on the *NewsHour* programme that night and the following morning. The Voice of America also carried it fully. I heard myself described as "renowned writer and environmentalist". Very supportive, I think, and I hope it goes on that way. When do you return?

Best wishes. And let me hear from you.

Ken.

2. Since 1994, he has also been Chairman of Shell Nigeria.

March 21, 1995

*Maynooth University Ken
Saro-Wiwa Archive.*

21/3/95

Dear Sr. M,

Seems like ages ago since I last heard from you. I have seen your work &
your pictures in the *Irish Times* though and I think you yourself might be surprised how far these Ogoni bells are ringing now, and how you have become
the bellman? I thank God for your presence among us. Ledum keeps asking
if you'll be able to return to Nigeria. And I say "yes" because I'm certain that
the present regime will fall and it will also be because people like you did so
much to alert the world to the danger it has been to the ordinary people of
Nigeria.

I have never really felt I was in danger. The sure knowledge of my inno-
cence gave me that feeling. I thought that I'd remain in captivity until God
should have used that fact to make the Ogoni cause better known and
pave the way for solving some of the many problems which confront the
Ogoni people and similar groups in Nigeria, if not the African continent. Big
thought. Big assignment. But I have not wavered in my belief. Now, seeing
how things are going, I'm even stronger in my belief and in my faith in the
ultimate success of my dreams.

It would appear that the British and the Americans are at last waking up to the danger posed by Abacha and his men. Really, Nigeria has not had it so bad. The denigration of the people is unimaginable. We have to be thankful that Wole Soyinka and his friends went abroad to alert the rest of the world. And that people like you are also out there to bear testimony independently. Ogoni bells. I'm in good shape. A touch of the flu now and again but otherwise, nothing serious on the health side which is a good thing. The "trial" is on, although no one is in doubt that the judgement has been written. What will ruin their plan is the fall of the Abacha regime, although one has to be careful then what Shell does next. I did let you know that Shell are represented by Counsel holding a "watching brief" at the trial. When that got embarrassing, they got the Counsel who is also Chairman of the Rivers State branch of the Nigerian Bar Association to publicly withdraw from the Tribunal while secretly representing them through the Bar Association which can legitimately be at the Tribunal.

The one problem I'm now facing is financial. I've had to take responsibility for everyone of the 36 people held. And also the MOSOP men and women who are underground. Gani Fawehinmi and his team fly into Port Harcourt for the trials and have to be accommodated in a hotel. And that has increased costs dramatically. He is not charging fees but there are two other Port Harcourt lawyers who are charging fees. One cannot object to that seeing that the Tribunal is so time-consuming. I did obtain permission from the Right Livelihood Award Foundation to use the prize money for our defence. I regret this a lot as I had hoped to spend the money on Ogoni women. The annoying thing is that it does appear that all the Prize money will be spent, along with even the Goldman Environmental Prize money (47,000 USD). The fact of these prizes has sent everyone to my doorsteps and imagination is running wild as to what money I can spend on even the most minor of errands etc. I feel terrible about this sometimes. I have sent word abroad that we need funds for our legal defence so it's possible either Glen or my son will have gotten in touch with you in this regard. If you can assist to raise funds, it will help.

You'd be amazed at the level of harassment and extortion that Ogoni people are facing at the hands of the military. The point is that the Internal Security Force is an illegal body. So there is no limit to what it can or cannot do. And when Nigerians get this sort of carte blanche, it is sure to drive them to bestial levels. The surprising thing is that the Ogoni are holding out so well. We need funds to continue to organise them for the resistance and to shore up their morale. As usual, the women have been wonderful.

I should also mention that a plus factor of the trial is the presence of Ledum Mitee who is not only a very good lawyer, but knows the story better than anyone else except perhaps myself. He was able to rubbish the testimony of Dr G.B. Leton[1] and Miss Priscilla Vikue[2] and to show them up as liars. Two other witnesses have also sworn to affidavits confessing that they were bribed to testify against me. In normal circumstances, that should have ended the trial. But ours is an extraordinary "Nigerian" situation and so the hangman must wait at the door for the man whose neck is wanted! Last week, the same charges were filed in two batches of 5 each against the other detainees. Each batch of 5 was accused of having committed the murders at the instigation of "Ken Saro-Wiwa, President of the Ogoni nation". You haven't seen such a farce. My lawyers insisted that I should be included in the charge and the prosecution found themselves in a quandary. Since there are going to be six batches, it would mean that I would be charged 7 times for the same "offence"! Indeed, under the law, I could be found guilty behind my back, if any of the suspects is found guilty. Besides, the same Tribunal could not try me and also all the others, using the same prosecution witnesses. Now we'll see how the Nigerian in these judges gets over these legal obstacles. I'm really in for the belly laughs! If only it were not so tragic.

I find the proceedings of the Tribunal boring or farcical or tedious. Often, I read newspapers. Sometimes I write letters – I actually was caught by Mr Mitee writing to Hauwa. A love letter when you are in a Tribunal facing a death sentence? It seemed incredible. He has been teasing me no end. Yet I do feel contempt for the Tribunal and there's no better way to express it than what I've done. I'd rather not have appeared before it. That was my gut feeling from the beginning, but it was felt that the lives of younger people were at stake and that we should give them a fighting chance.

We challenged the jurisdiction of the Tribunal. Under the law, there should have been an Investigative Committee preceding the setting-up of the Tribunal. This was not done. Yet the Tribunal felt it had a right to proceed.

In rejecting our application for bail, the Tribunal actually said that we had killed the deceased. This is going to serve as a basis for challenging the right of the Tribunal members to continue the trial at all. I expect they will also

1. See the letter dated 30/7/94, n. 26.
2. Director General of the Ministry of Education in Port Harcourt.

throw out the motion. Then we'll go on to make a "No Case Submission" after the prosecution shall have concluded their case.

And so the farce goes on. I'm trying to perfect some of the manuscripts I've completed during my incarceration. The pity is my computer has been seized and my work is therefore stymied. However, I hope to complete the diary of my first detention and to send it off to the U.K. in the hope that I might find a publisher. Also a collection of short stories A *Kind of Festival & Other Stories* which I believe to be the best of the three collections I've done so far. I don't know if my new-found "fame" or notoriety translates to the sale of my books. Anyway, my books are badly distributed in the U.K. and elsewhere. I'm not so sure that I should have allowed Longman to reissue *Sozaboy* and A *Forest of Flowers* in their African Writers Series. They tend not to distribute well and their advertising is rather below the mark while their main interest is in educational publishing.

I don't know if I've told you about Shell's "Major Environmental Survey" of the Niger Delta for which they've put up 2 million USD. How that amount can conduct a "major" survey is to be seen. Gamaliel Onosode is said to be the Chairman of the team in charge of the survey. He has said some good things about me in the past. I wrote him, including *The Drilling Fields* and my other publications and asking him not to allow himself to be used by Shell. I'm yet to hear from him. But Owens will be looking him up next week.

Richard Boele of UNPO was here and has done a report which should be of interest to you. African Watch[3] were also down here. One of them, Melissa,[4] is said to have stumbled on a report by military intelligence absolving me of all suspicion. Melissa took up the matter with Philip Umeadi, the prosecutor. The latter is reported to have said, "Nigeria is a primitive country" so innocent people can be charged before a kangaroo court and sentenced to death.

I expect I've done a run-down of the situation as it is. Oh by the way, I understand Bodyshop have given us fifteen thousand pounds for our defence. A group of Nigerians in Australian have also sent a token sum.

3. Human Rights Watch Africa: Human Rights Watch (HRW) is a nongovernmental organization which monitors and promotes the observance of internationally recognized human rights in Africa, the Americas, Asia, the Middle East and among the signatories of the Helsinki accords.
4. Probably Melissa Crow who was a Sophie Silberberg Fellow at HRW from 1994-95.

The Goldman Environmental Prize[5] is to be awarded April 17. I'll be sending Meschack Karanwi who handles MOSOP publicity to represent me. I hope that Ken Jnr. can also go down there at my cost. I hope that some publicity follows it in the U.S. and helps to embarrass the Abacha Govt.

I hope that your health is holding out and that we'll be able to see you again in due course.

God bless you.

Ken.

---

5. The world's largest prize honouring grassroots environmentalists. Founded in the U.S.A., the recipients are selected by an international jury. The prize money amounted to 47,000 USD.

March 24, 1995

Sis M,

Just a quick note to say I've got your diary of 7th Feb—45 days late! But it's brought joy to my heart, as usual. I've read it twice now. Sent me back to the short story "The Resurrection" —not one of my best; I wonder what you find in it. My brother was at St. Patrick's Day in Lagos and found time to speak with a number of ambassadors, he tells me. The French Ambassador is reading *Darkling Plain*. I'm trying to perfect my first Detention Diary *A Month and a Day* and I've completed the corrections on my latest short story collection *A Kind of Festival and Other Stories*. I think this collection my best so far. I'll be sending both to Junior Ken & asking him to see if he can get a publisher in the U.K. My newly-acquired fame might just help.

Did I mention the Goldman Environmental Prize in my earlier letter? The Hunger Project[1] wrote me about this year's prize. Could you nominate me again? My particulars are already on file. All they need is a letter before May 15 re-nominating me. Thanks.

Keep smiling. God loves you.

Ken.

---

1. A global non-profit organization which works to promote the sustainable end of world hunger. See <http://www.thp.org>.

May 2, 1995

2/5/95

Dear Sr. M,

Greetings. I hope you are well. It's taken some time since I sent you letters through Julia. I expect you will have received them. Somehow, I feel I'm going to hear from you in the next few days but since Claude Ake is travelling, I'm seizing the opportunity to brief you as to recent events.

I'm supposed to be in hospital—the military hospital here in Port Harcourt. The Tribunal made the ruling on the 28th. Then it transpired that the place was in a mess and had to be painted and equipped to receive me. The doctor in charge, Lt-Col Agada (he's a Catholic) came to see me yesterday and confirmed that the room is ready. We are now waiting for Lt-Col Okuntimo to release me. I hope I can find rest there if nothing else for I'm mentally exhausted.

That apart, I'm in good spirits. My brother Owens has had a breakthrough in Lagos. He had an hour-long recorded conversation with the British High Commissioner who assured him Shell was on the side of the Ogoni. Thereafter, the High Commissioner arranged a 45-minute discussion for him with Brian Anderson. Brian said he did not trust me although he recognised that I had the full support of the Ogoni people. Then he asked what he could do for us. My brother asked him to get the Tribunal stopped. He said he couldn't do that; that Abacha's owing Shell a lot of money which he refuses to pay. Then my brother raised the matter of my being sent to hospital. He thought he could do that. The Managing Director of Mobil and the British Deputy High Commissioner were present in each half of the conversation. To cut a long story short, Brian gave Owens his office and home telephone numbers, thus signalling they could hold further discussions. I've since given Owens the go-ahead for further talks and asked him to ensure Mitee & I are out of this place since we are the only ones who can negotiate anything on behalf of the Ogoni people.

Shell have now gotten their decision, declared in their recent paper *The Ogoni Issue* widely publicised in Nigeria: that they will not return to Ogoni until & when the local community is in harmony with them. I hope that that gives the requisite signal to the Government—that they have to accede to the demands of the oil-bearing areas and not visit them with violence in order to silence them. Because silencing them does not mean that oil production can go on as usual. I hope also that other oil-bearing areas are listening. If they are, that should introduce a new situation into the Nigerian equation. That is if they can organize as the Ogoni have done to demand their rights in a peaceful manner.

At the local level, there are moves towards reconciliation as everyone seems tired of the problem. When Dr Agada (Lt-Col) came to see me yesterday, he asked if I do pray. He said I should have more faith in God than in the legal. He has access to both the Military Administrator and the Chairman of the Tribunal and both did not seem to be seeing their way through the Tribunal. The Mil. Administrator thought he had been acting to save his administration which we were threatening. But then he had found no peace this year. He appeared to be wanting some face-saving way out, more so as his tenure seems to be coming to an end. I advised that Lt-Col Agada get Mr. Mitee to speak to the Chairman of the Tribunal and the Military Administrator so a way out can be devised. Whether anything comes of that remains to be seen. But if I get into the military hospital, I'll be seeing Dr Agada quite a lot and may be able to nudge him on.

In Ogoni itself, the vulture chiefs seem to be wanting a resolution of the problem. They've made overtures to us and some of our men are beginning to speak to them. Even the prosecution witnesses want to be paid to alter their testimony same as they were paid to bear false witness. Ledum Mitee is handling that.

I doubt that my being allowed to go to the hospital owes itself to Shell. Before that, we had made all arrangements ourselves with the doctor from the Teaching Hospital who was sent to examine me. Whatever the case, I'm sure that the pressure from abroad is biting. With the Goldman Prize, a lot of publicity has been generated in America. The Human Rights Watch Africa Report was very damning and Kenneth Roth, Executive Director of HRWA [Human Rights Watch Africa] did write an open letter to Abacha asking him to allow me medical care, disband the Tribunal and set all the Ogonis now held free. I expect that the British QC, Mr Birnbaum[1] who watched the trial for a week will have made a report to the British Law Society. The British High Commissioner has said that his report could influence the British Govt. And he left us in no doubt that he did not think the trial fair.

---

1. Michael Birnbaum, a British lawyer appointed to the Queen's Counsel in 1992.

My son has been in Canada and I understand he had some airtime on CNN during his recent trip to America to receive my prize. It's been good for him. I heard him on Voice of America and he sounded quite good.

All in all, the outlook seems hopeful. The suffering in Ogoni is quite depressing. The goon soldiers are still around and are still extorting money from people. And at the national level, the economy is in a big mess. Abacha is still playing tough and stupidly so. The people cannot dislodge him; only foreign pressure will do that. I'm hoping that the TransAfrica initiative from the US bears fruit. In any case, we live in expectation of the collapse of the govt. one of these days.

I must stop here until I hear from you again. My warmest wishes.

Ken.

June 19, 1995

Maynooth University Ken
Saro-Wiwa Archive.

19/6/95

Dear Sr. M,

Just a quick note to confirm that I received your letter of late May. And
many thanks. I'm having to get used to the fact that you will not be returning
to Nigeria and that I may not see you for some time. I hope that it does not
take too long before we can meet again.

The contact with Shell has dried up, thanks to the very useful and necessary demonstration at the Shell AGM of 18th May. I hope it resumes, since
we need to dialogue. I know they will do everything to resist us and that
they may still want me out of the way. I am not careless of my safety, but I
do recognize and have always recognized that my cause could lead to death.
But as the saying goes, how can man die better/than facing fearful odds/
for the ashes of his fathers/and the temple of his Gods? No, one cannot
allow the fear of death to dent one's beliefs and actions. I only wish there
were more Ogoni people on the ground. However, the cause cannot now die.
There are people like you, Anita Roddick and even my son Junior has seized
the gauntlet. Then there are the resilient Ogoni on the ground. You should
see how Ogoni women have seized the gauntlet. My mother supervises a
weekly meeting of Ogoni women from all of 40 villages (nearby ones) and

they are really a joy. She tells me that to escape military attention, they operate under the banner of the National Council of Women's Societies [NCWS]. (A few years back, she was honoured by the NCWS as one of seven women achievers in Rivers State).

The so-called trial is proceeding slowly. Our lawyers should be pulling out any day now, since the whole thing is a charade and the Tribunal Chairman has left no one in doubt that he has written the judgement in advance. We'll see how things proceed thereafter. I see them shutting me up for the period Abacha is in power. But I'll be stronger in prison than out of it. It's a price I'm not afraid to pay.

I've done a few poems recently. They're still in the rough but you may savour them.

All my love,

Ken Saro-Wiwa.

July 8, 1995

8th July, 1995.

Dear Sr. M,

I got your diary of 5/5 to 16/6, and many thanks. As you never tell me which of my letters you have got and which you have not, I can only hope that you have got all the letters I have sent you. Like you, I only hope that my letters do arrive.

I did get the book, GETTING TO YES,[1] and many thanks for it. You must have known that the exploratory talks with Shell petered out even before they began. I have a feeling that Shell thinks they can sit all this out and carry on as before. I think they are wrong. All it means is that we have to work harder. It may even work in our favour that we do not negotiate just now, but that the situation in the country goes bad enough to make a total resolution of the contradictions necessary. In which case, we will also be negotiating for other peoples who are not up to it at this moment.

You know of the Niger Delta Environmental Survey. Claude Ake sits on its Steering Committee, representing the "stakeholders" i.e. the local communities. He is constantly in touch with us, and has, indeed, held two meetings with us here in the detention camp. He confirms what we all knew: that Shell had wanted to do a "greenwash". However, it does appear that they won't be getting their way, it being that Claude is very clear as to what should happen, and is receiving support from one other member of the Committee–the chap from the World Conservation Fund who comes in from London and who is determined to ensure that his organization is not misused by Shell. There is a stakeholders meeting in Port Harcourt, 24th to 26th August and I will let you know what goes on there. In the meantime, could you please check with Glen if it's true that the Head of the Environmental Division of Shell at the Rotterdam office resigned rather than be a part of the grand deception which Shell planned in instituting the survey? The position paper appears to have sought for a way of using the survey to create communal dissension and for buying over local leaders. Check this out please.

1. Roger L. Fischer, William L. Ury, Bruce Paton, *Getting to Yes: Negotiating Agreement Without Giving In* (London: Houghton Mifflin, 1981.; London: Penguin, 1982).

Just as you were leaving for Rome, Dr Leton and Dr Birabi were flown to London by the federal government to try to undo some of the publicity the Ogoni campaign had put out. It would appear that Shell hosted them to dinner but I cannot say that they met with much more success than that. Ojukwu[2] was on that same trip but I understand that they were booed when they spoke to their first audience and their first press conference was similarly a failure. The behaviour of the Ogoni politicians has been roundly condemned at home.

The trial has been adjourned to July 31. Our lawyers have now withdrawn and the Legal Aid Council of Rivers State has provided a lawyer for each of us.[3] They now need to obtain all the proceedings and study them before the trial can recommence. We expect that they will get the proceedings from the Tribunal but we are not giving them any of our own papers. In short, we will not be co-operating with any of the defence counsel. We would like it known that the State are judge, prosecuting and defence counsel all rolled into one so that the true intent of the State is no longer masked.

Our lawyers withdrew after the Tribunal ruled that the failure of government to produce a video-cassette and transcripts of a press conference addressed by the military administrator and one of the prosecuting witnesses a day after the murders, was "reasonable". The government said they had wiped all the tapes for lack of funds. But they had in fact distributed the same throughout Nigeria and their embassies abroad in an attempt to calumniate me. So much for the trials.

The situation around us in the detention camp is a lot more relaxed now. "Delta Force" forced Okuntimo out of his positions as Camp Commandant and Commander of the Internal Security Task Force and he has now been

2. Chukwuemeka Odumegwu Ojukwu, military governor of the Eastern region who led the break-away Republic of Biafra during the Nigerian Civil War 1967-1970. At the end of the war, he received political asylum in Côte d'Ivoire but returned to Nigeria in 1982. Although imprisoned for a period between 1983-84, he was eventually released and held various political roles in Nigeria until his death in 2011. Saro-Wiwa characterized him as misguided and opportunistic in his memoir, *On a Darkling Plain: An Account of the Nigerian Civil War* (London; Lagos; Port Harcourt: Saros International Publishers, 1989).

3. The lawyers for MOSOP withdrew because of the unconstitutional way in which the trial was being conducted. The state then appointed lawyers to each of the nine accused but they decided not to co-operate.

thrown to a harsh position in Minna. Consequently, we are now in ostensibly more civilized hands, although this is only relatively speaking. For a greasing of palms (you cannot avoid that in Nigeria— everyone's trading), people have access to us and can spend thirty minutes to one hour. This has been most relieving.

We have met with the youths and the women of Ogoni, and have been able to get more accurate pictures of what is happening at home. The military harassment has been most frightening, and most people are scared to death. It has been impossible to hold meetings, and it does appear that the only meeting that is currently being held once a month is that summoned by my mother. She is patron of the organization—BIAKA (GLORIOUS WOMEN)—and the last meeting was attended by 200 women from all over Ogoni. The next meeting holds on August 1 and I will be arranging for a video of that. I will also encourage them to watch *Delta Force* and *Drilling Fields*. They are surely in the vanguard of the resistance. I feel so proud of my mother! And she does it all so naturally! She's always at the Tribunal. So also my father who spent more than an hour with me here in prison a few days ago.

The men and the youth who seem to be the target of the military are more wary, but I expect that now that they have access to us, we will be able to motivate them. It is sad that all those whom we were training are now underground or in detention. This has left the people leaderless and am worried that should this go on much longer, we may lose momentum. However, we are now trying to encourage a new leadership, but this is not going to be easy. The harsh economic condition has placed too many burdens on people and resistance in the circumstances, especially where government is so very powerful, is a clear difficulty. We need a lot of money to organize the people, support those who are underground or are detained and pay legal fees. I must tell you that I've been under considerable pressure but I'm soldiering on with great faith in God and in my ultimate destiny.

I believe that a settlement will come with the resolution of the national crisis. In that sense, the recent coup trials which have outraged the international community are welcome. The point, though, is will the world watch hands akimbo as they did in Idi Amin's Uganda while the best men are murdered? I fear that that might be the case, as the people in government who should act on these things are normally moved by every consideration but the preservation of human rights. More so when the crimes are being committed against people in the Third World and commercial and industrial interests are involved. Zaire comes readily to mind.

I did receive a laptop computer from Junior through my younger Sr. [sister], Comfort, the Zaria-based lawyer who was in Britain for a fortnight or so. This has eased my pain a great deal, and am beginning to write again, after a lull of seven months. Seven wasted months! I'm encouraging Ledum and John Kpuinen[4] to also learn how to use the computer so they can keep themselves busy.

You will be telling me about your month-long stay in Rome in your next letter, no doubt. And also let me know how you settled into your new situation. I can't just reconcile myself to the fact that we will not be seeing you in Ogoni again, at least not in the foreseeable future. But then, I'm not in Ogoni either. What is comforting is that you are well and that you will always be in touch, quite apart from the fact that spiritually, we are together.

I got the photographs of the march, and was very pleased to see your seraphic person looking tense in one or two of them.[5] Some lawyer from Ledum's chambers grabbed them to make photocopies, so as I write, I've not really had time to look at them closely. I'll do so in due course.

Please give my regards to members of Ogoni Solidarity Ireland and say how grateful we are for their care and concern. Also give my thanks to Mairéad Corrigan.[6] I cannot believe that anyone should think me fit to receive the Nobel Prize for Peace. I also hate to think that I've not been able to travel to receive any of the prizes I've won save the Fonlon-Nichols award.[7] The joy lies in the recognition. To think that I should be the Nigerian with the greatest number of international awards. And that I should be an Ogoni! What's

4. Deputy President of MOSOP's youth wing, the National Youth Council of Ogoni People (NYCOP). He was among the "Ogoni nine" who were hanged in November 1995.
5. Most likely, this refers to the annual Afri Famine Walk which McCarron attended in Mayo, Ireland, in 1995. Afri is an Irish Non-governmental Organization whose activities revolve around human rights and peace and justice issues. Photographs of this walk are held in the Ken Saro-Wiwa Collection at the National University of Ireland Maynooth.
6. A Northern Irish peace activist and co-recipient with Betty Williams of the Nobel Prize for Peace in 1976. The two women co-founded Community of Peace People, an organization dedicated to finding a peaceful resolution to the conflict in Northern Ireland. Corrigan nominated Saro-Wiwa for the Nobel Prize for Peace in 1995.
7. An award conferred annually by the U.S.-based African Literature Association for work promoting human rights.

that about the stone rejected by the builders? God is good, as we say over here.

I must stop here and wish you the best of luck. I look forward to hearing from you soon.

Sincerely,

Ken.

September 14, 1995: Ken Saro-Wiwa's last letter

Dear Sr. Majella,[1]

Many thanks for your letters up to the end of July or thereabouts. I believe that I've got everything you have sent thus far. Some of them come rather late and out of sequence, but I do get them. Because I keep them around me just to read and re-read them, I've had two of them seized lately.

I hope that I will get them back, anyway, some day.

I expect that you have now started your new assignment and am really happy for you. It is hard to think that you will no longer be with us here in Nigeria, but it may well be that we shall be better served by your being away. God works in mysterious ways his wonders to perform.

You have heard of the raid on my office of July 29. That was followed by a raid on our cells in the prison a week or so later. Since the departure of Okuntimo on July 1, and indeed long before then, for Okuntimo suddenly lost interest in punishing us, thanks to the education we gave him, security had eased a great deal and we were able to receive a host of visitors from Ogoni and elsewhere.[2] I'm sorry to say that we became lax ourselves and so when the new man struck, we "lost" quite a bit of property.

Back to business. The trial is being speeded up. Which means that every rule is being broken in order that it might end. I expect that it will end before you get this letter. I have information that judgement will not be delivered, which will mean that we will remain in prison for as long as the authorities please, keeping me out of circulation for as long as Abacha is in power.

I am not defending myself. I will want to make a Statement. I have already filed the Statement at the High Court here and a copy is available with my son in London, if you want to read it. It is possible that the Judge will not allow me to read it, but it is already a public document and the press can use it.

About a week ago, Major Obi Umahi (an Igbo) who now commands the ISTF

---

1. This is the last recorded letter written by Ken Saro-Wiwa. It was hand delivered to McCarron in Belfast by Saro-Wiwa's eldest son, Ken Wiwa, after his death.
2. Okuntimo was transferred to another area

[Internal Security Task Force] came to me and asked what he could do to ensure peace in Ogoni. I asked him to release the three Ogoni activists who were arrested after the raid on my office, and use them to speak to the Ogoni people. He promptly released them from detention at Afam the next day and brought them to my cell. He then sought my blessing of the peace effort he meant to launch. I gave that blessing after drawing his attention to the fact that there are three parties involved in the matter: the Ogoni people, Shell and the Government (State and Federal). I did not mind where he started from, I said, so long as he realized that peace could only come when justice is done.

As of now, a core group of "moderates" is being put together, preparatory to the group being widened to include the disaffected politicians. The hope, I gather, is that the Ogoni people can be made to come together to defend Ogoni interests. I don't know that anyone can reconcile the skulduggery of Leton and Birabi with the rest of the people. My assumption is that Government are preparing the ground for the conviction of Ledum and myself—to ensure that the Ogoni do not riot when the inevitable verdict, long decided upon, is handed down.

Yesterday, the ground was cleared for that conviction when the no-case submission made by the Legal Aid lawyers was dismissed. The Tribunal said: "The prosecution adduced evidence of the killing of 4 chiefs of Gokana at Giokoo through civil disturbances there by NYCOP youths.[3] There is evidence of the leadership of MOSOP who by meetings, rallies, provocative and instigative pronouncements light(sic) the fuse that produced the consuming events of 21/5/94 at Gbenemene Palace, Giokoo." Whoever was in doubt knows where the Tribunal is heading to. The stories below are for your entertainment.

By July 1, Okuntimo and Komo were at daggers drawn. I believe Komo finally saw through the brute he had appointed to supervise the decimation of the Ogoni and knew that he had made costly mistakes. As usual with the military, he could not apologize to us; in any case, his career was already on the line. All he could do was persist in evil.

The Federal Government were shamed at the last meeting of the UN Committee for the Eradication of Racial Discrimination (CERD) attended on our

---

3. The National Youth Council of Ogoni People. Saro-Wiwa had prompted the formation of NYCOP as well as other grassroots organizations.

behalf by Barika and on the Government side by a team which included George Kobani (as consultant), the Legal Adviser to Abacha, a Professor Yadudu (Yahoo!) and Ambassador Azikiwe (second cousin of the great Zik[4]). The Government delegation declared me guilty of murder (!) and the Committee felt that since they were denying all the assertions made by the various human rights groups, they might as well go to Ogoni to see things for themselves. I doubt that they will be coming, but I wish it would happen.

A newspaper here published an item stating that Shell has been pressuring the government to create an Ogoni State and give the oil communities 50% of oil revenues. Shell has not denied this item, and there may be some truth in it.

Ogoni is buzzing with all sorts of noise, all sorts of expectation. The first meeting aimed at reconciliation was held yesterday (13th) the date on which our defence opened. I believe that the government will be told that peace in Ogoni depends also on all those now held in detention or driven underground being free. They won't listen to that. I can only hope that our conviction does not lead to more trouble. I have tried to bring home to all the example of Christ which was, to our people, fictional, I guess. They are now going to learn a hard lesson and I hope they take it quietly.

I will keep you posted as events unravel. The ways of this struggle have been so unpredictable, I don't know what to say. There is an unseen hand at work. We are just not in control, and no one seems to be. It's amazing the things which do happen!

I must thank you and all those who have done so much work for us abroad. It's nice to know that people who have worked or visited Ogoni are now turning up and lending assistance in various ways. The commitment of Anita Roddick is a great blessing. It has nettled the Nigerian High Commissioner in London, for one!

I am in good spirits, expecting the worst as usual, but hopeful for the best. My parents are always in court, and my father believes that I will be free at

---

4. Nnamdi Azikiwe, one of the leading figures of anti-colonial Nigerian nationalism. Prior to independence, he led the National Council of Nigeria and the Cameroons—a party of the Eastern region with a primarily Christian support-base which espoused Pan-Africanist principles. He was the first independent President of Nigeria from 1963-66.

the end of the case. I've tried very hard to dampen his optimism but the old man won't budge. I just hope he does not get a rude shock.

One source of worry is what will happen to our struggle when Ledum and I are put away. We had not had enough time to train the cadres or put alternative leaderships in place. And putting members of the Steering Committee on the police wanted list has deprived us of a lot of hands. I have been able to direct things and even contribute to the publicity war from detention. I don't know if I'll be able to do so from prison. We have no funds, not even a bank account. Everything had hinged so much upon my resources that my absence will cause a lot of problems. We'll have to get around that somehow.

Don't be embarrassed by my ocean of ink. I am not. I'm only tired now and must go to bed, to rise early and prepare for the boredom of the Tribunal.

Regards

Ken

POEMS

Ken Saro-Wiwa. Courtesy of Noo Saro-Wiwa. Maynooth University Ken Saro-Wiwa Archive.

I lie alone at night

I lie alone at night
And think all of one year's gone
Since I held you in my arms
In the bed we know so well.

I lie alone at night
And see the callous bandits
Break into our hallowed bedroom
Cruelly knife our togetherness.

I lie alone at night
And think of you lying lonely
Dreaming of my return
To the home we love so well

I lie alone at night
And think of the thick boots
Which stalk the halls of tyranny
And crush us underfoot.

I lie alone at night
And wonder why you wait
And endure the gripping pain
Which is my lot to bear.

I lie alone at night
And think of the stranger moon
The stars beyond my gaze
Your beauty like moons and stars

I lie alone at night
And pray the day will come
To mend your broken heart
And steel my breaking soul.

I lie alone at night
And dream a great new dawn

Without boots and knives
Broken hearts, breaking souls
Empty dreams and lonely beds
Stranger moons and searing pain
When you and I and all of us
Can hold hands and sing our love
Into a night captured by peace.

Ken Saro-Wiwa
3/6/95

For Anita Roddick

Had I a voice
I would sing your song
Had I a tongue
I would speak your praise
Had I the time
I would live for you
But here, gagged and bound
To the floor and injustice
And waiting for death
I can only wonder
How you have breathed life
Into the thought that we give
Meaning to being
Only when we share and care.

Ken Saro-Wiwa
20/6/95

For Sr. Majella McCarron

Sr. M, my sweet soul Sr.,
What is it, I often ask, unites
County Fermanagh and Ogoni?
Ah, well, it must be the agony,
The hunger for justice and peace
Which married our memories
To a journey of faith.
How many hours have we shared
And what oceans of ink poured
From fearful hearts beating together
For the voiceless of the earth!
Now, separated by the mighty ocean
And strange lands, we pour forth
Prayers, purpose and pride
Laud the integrity of ideals
Hopefully reach out to the grassroots
Of your Ogoni, my Fermanagh.

Ken Saro-Wiwa
20/6/95

The celebrants

They are met once again
To beat drums of confusion
Tattooes of mediocrity
They are met once again
The new cow to lead
To the cyclic slaughterhouse
Where the blood of the last
Yet stinks to skies
Awash with the stench
Of decay and corruption.

So they come and go
Bad dreams, phantoms
In the morning mist
Meteors of a dark sky
Offal at the desecrated shrine
Still they come and go

We welcome them from rooftops
Raucous at each approach
And departure, praise-singing now
Cursing loudly later
Impious votaries of an alien sect
At whom damnation shrieks with wild delight.

Fire

There is a fire in me
Burns all night and day
Flares at injustice
Leaps at oppression
Glows warmly in beauty.

Prison song

Bedbugs, fleas and insects
The howl of deranged suspects
The dark night bisect
Rudely breaking my nightmare
And now widely awake
I'm reminded of this crude place
Shared with unusual inmates.

Mama came calling

She came visiting today
The lovely little lady
In her hand a dainty meal
Of nutless palm fruits
A long-forgotten delicacy
From my childhood days
Into which I dug my teeth
As my baby gums her breasts
And found therein once again
The milky sweet of a mother's blessings.

Victory song

You have raped my land
Black brother, silenced my song
Upon my wholesome breath–
Condemned to a gas dungeon
I suffocate, shriek in pain
Into cold, stone-stuffed ears.
Your fingers drip with my blood
Staining your nails black and crude.
Vampire, tyrant, rapist
Black brother of the same womb
But cruel as the flares that burn
Poisonous gases into our skies.

I lie manacled in chain
In caves of your callous care
But the day will come when I will
break your hard bones
With my claws tear your brain
Consume you in wrathful fires
To the wild winds expose you
Paint the cruel marks of your sin
On the walls of history.

Then shall I, triumphant
Return to our hapless mother
With bright bouquets of peace.

Ogoni! Ogoni!

Ogoni is the land
The people, Ogoni
The agony of trees dying
In ancestral farmlands
Streams polluted weeping
Filth into murky rivers
It is the poisoned air
Coursing the luckless lungs
Of dying children
Ogoni is the dream
Breaking the looping chain
Around the drooping neck of a shell-shocked land.

MOSOP

Mosop is marching on
Ogoni must survive
Mosop will never stop
Till Ogoni is free

Mosop keeps marching on
Mosop is an arm of peace
Mosop stands for justice
Mosop will never stop
Till Ogoni is free
Mosop keeps marching on!

Mosop is for hard work
Mosop stands for success
Mosop will never stop
Till Ogoni is free
Mosop keeps marching on!

Mosop is for our land
Clean water and pure air
Mosop will never stop Till Ogoni is free
Mosop is marching on!

Mosop is all our life
Ogoni must survive
Mosop will never stop
Till Ogoni is free
Mosop goes marching home!

Ogoni hymn

Creator of Ogoni
Land of glory and wealth
Grant us thy peace and lasting love
Plant justice over our land
Give us thy wisdom and the strength
To shame our enemies.

Creator of Ogoni
Land of glory and wealth
Grant everlasting blessings Lord
To people of Gokana
Khana, Eleme, Tai and Babbe
Glorious Ogoniland.

Geneva

Two rivers
And a lake
Shimmering
In summer sun
Glittering
Antiseptic
Twixt hills
And prices high
Hosting nations hostage.

Around the court of Abuja fame

Around the court of Abuja fame
Sit grand professors we dare not name
So clever they in politics
They sure do truck a lode of tricks.

The praise of their deaf listening king
From morning to cursed night they sing
Blaming hapless politicians
For cheating on their ambitions.

Around the court of Abuja fame
Propounding great theories crude and lame
Our professors rant chant and dance
Their rank ceevees they do enhance.

Around the court of Abuja fame
National service is just a game
They mouth and dribble all about
While people starve and die and rot.

Babangidance (1)

A little to the right
A little to the left
Watch your step shake your waist
Dance to time watch your tilt
Don't be slow don't be fast
Sing a song watch its lilt

A little to the left
A little to the right
A bit of that a bit of this
Bite a bait but don't be bit
Don't dare demand what it is
Makes Babangidance such a hit!

Babangidance (2)

A little to the left
And nothing can be right
One step forward
And five backward
And stand at ease
Babangidance.

Charred woman

The boiling hale of gas fumes
Have burnt Zora's hand and body
The callous hand of cruel neglect
Her wretched soul destroyed
She's mere ash and burning misery
Charcoal of human greed.

For Zina

I have raised the questions, daughter
Which you and your kids must ponder
I feel guilty I did not sooner
In my lifetime urge them stronger
And now, ere I answers provide
I may in cold blood lie buried
Have I your futures compromised?

Stone deaf

Tired
of the questions
bored
by the din
of the lone voice
piping
to ears hearing
only
the sounds of their choice.

For Tedum (B. 8.7.78) who died at Eton College (15.3.93)

But one brief flash and he was gone
The young tree, growing e're so fast
Planted in the joy of summer
Watered through a mellow autumn
And cut in a brutish field
In the harshness of despairing winter
Lending a personal meaning altogether
To Shakespeare's ides of march.

Oh my lovely young oak,
Could I but see you once again
E'en in my dreams forlorn
Life would become a season of blossoms
Beautiful, promising and sensuous
Enlivening this languishing prison.

Morning song

This morning is sheer poetry
as from my detention cell
my heart sings with the red
freshness of hibiscus flowers
the vivid colour of the ixoras
shooting out of the green abundance
of a heart which resists surrender
to a garden of rank weed and mush.

Whore

The midnight knock on my bedroom door
Has turned me into a notorious whore
Open to every ugly steel penis
Pouring forth bullets of solid sperm
The sight of which I dare not resist
Lest I earn the right to choose my lover.

Town-crier

Take these cuffs from my legs
And set me free
Pick the lies from your teeth
And let me be
Town-crier, proud gong
Calling the lame and deaf
To defend their blasted land.

Night time

The breep of insects
The hoo-hoo of bullfrogs
And the croak of toads
Companion of a night
When nightmares burgle our sleep.

The call

Hear the call of the ravaged land
The raucous cry of famished earth
The dull dirge of the poisoned air
The piteous wail of sludged streams
Hear, oh, hear!
Stunted crops fast decay
Fishes die and float away
Butterflies lose wing and fall
Nature succumbs to th'ecological war.

Keep out of prison

'Keep out of prison,' he wrote
'Don't get arrested anymore.'
But while the land is ravaged
And our pure air poisoned
When streams choke with pollution
Silence would be treason
Punishable by a term in prison.

Detention haircut

I had a haircut today
After a long delay
My prison hair had grown so long
I thought it was full of lice
It looked thoroughly unkempt
A barber my jailers refused
So I did it just my way.

And oh, when I was done
You'd think I'd been visited
By a carpenter rude and crude
Or that an army of mice
Had raided my lovely head.

I could not stand the mirror
It told a tale of horror
But what I most feared
Was my aged mum would dream
I'd had this grisly haircut
As once before she had
And came upbraiding me
'Cause I looked like a convict
A disgrace to her proud womb.

On the visit of the doctors to my cell

The doctors are coming,
this odd, mystic morning
to cart our corpses to the ruling
magician at whose bland bidding
hospitals become clinics become
mortuaries and surgeons morticians
dancing with military politicians
to band hits of a cruel symphony
makes of a crude cacophony
finetuning physicians phoney.

Tall love

Nesting one tender morning
In cool, scented, pillowed sheets
I ran lost over your warm hips
Suddenly tear drops like petals
Cascaded down your flowery cheeks
And you were sad as a voyage.

Your sigh was the wind
Hurtling past dumb cliffs
On a troubled starless night
Still your arms of flowers wrapped me fondly round.
   Oh love, love, tall as the wind
Oh slender elegance of the palm
Let me sip the milky sweetness
Of your tender breasts
Then safely lost in the soft warmth
Of your mango laps I'll steep
All sorrows in love's sweet endeavours.

Ken Saro-Wiwa.

Afterword

MARK DUMMETT

TWENTY-FOUR YEARS AFTER the execution of Ken Saro-Wiwa and his eight colleagues (the Ogoni 9), I welcome the publication of a new edition of *Silence Would be Treason: Last Writings of Ken Saro-Wiwa*. These letters and poems portray the writer, the activist, the family man and above all a man who passionately cared about the manmade ecological disaster in his homeland. Writing to Sister Majella, in a letter dated 1 December 1993, Ken Saro-Wiwa wrote:

> *Keep putting your thoughts on paper. Who knows how we can use them in future. The Ogoni story will have to be told.*

The sad truth is that more than twenty years after the execution of the Ogoni 9, Nigeria's oil producing region remains a blighted land. There are hundreds of oil spills in the Niger Delta every year and Shell and the other oil companies operating there are still not doing enough to either prevent spills, or clean them up. The impact on the hundreds of thousands of people unfortunate enough to live next to the oil wells and pipelines where spills occur is catastrophic.

Shell, the largest operator, likes to blame local communities for the pollution, accusing them of cutting open the pipelines to steal oil. This is indeed a problem, but Shell overstates the issue to deflect criticism of its own failings, such as the poor state of its pipelines, and its terrible record on clean-up.

Ken Saro-Wiwa wrote that oil pollution had turned the Niger Delta into an "ecological disaster." The United Nations Environment Programme vindicated his claim – described by Shell at the time as false – in 2011. Its researchers found that the people of Ogoni, Saro-Wiwa's homeland, had "lived with chronic oil pollution throughout their lives." This pollution had contaminated the fields where they grew food, the water where they fished and the wells from which they drank.

Amnesty International campaigns for a proper clean-up of the Niger Delta because of this clear link between the oil pollution and the impact it has on the health and the livelihoods, and therefore the human rights, of the people. Talk to anyone over the age of 60 in the Niger Delta and they speak wistfully of swimming, in childhoold, in the clean waters of the creeks, which meander through the region – one of the world's most ecologically important wetlands.Visit these creeks today and you see signs warning people not to go close, you see dead mangrove trees lining the shore, and mud that is black with oil. The people are understandably angry, and have refused to allow Shell to pump any more oil from its wells in Ogoni.

But tragically, the pollution continues. Shell, which has been operating in Nigeria since the days of the British Empire, transports hundreds of thousands of barrels of oil a day along pipelines that cross the villages, fields and creeks of Ogoni from neighbouring oil fields. These pipelines are old and leaky. We know this because of internal Shell documents that the company was forced to disclose during a recent legal action in London. The court papers include an internal memo by Shell based on a 2002 study that states that, "the remaining life of most of the [Shell] Oil Trunklines is more or less non-existent or short, while some sections contain major risk and hazard." In another internal document dated 10 December 2009 a Shell employee warns that, "[the company] is corporately exposed as the pipelines in Ogoniland have not been maintained properly or integrity assessed for over 15 years."

In August 2015 I travelled to the village of Kegbara Dere, in Ogoni with colleagues from Amnesty International and the Centre for the Environment Human Rights and Development, which is based in the Niger Delta. There, we visited a place called the Bomu Manifold. It is an important facility for Shell that is guarded by the Nigerian military. The manifold is a hub for the company's pipelines, which run from the oil fields to an export terminal on the coast. In 2009, there was an operational fault on one of these pipelines, which then exploded, causing a major spill. Under Nigerian law, Shell has an obligation to start cleaning up spills within 24 hours, whatever the cause, and return the affected land to as close as possible its original state. The company said it had completed this work at the Bomu Manifold in 2012. Yet we found that a wide area was still visibly contaminated with oil. A large area of land looked like a bomb had landed on it. There was no vegetation, but mounds of charred and oil encrusted soil. We saw streams spreading the contamination into a wider area, where people live and farm.

We visited another three locations which Shell said it had cleaned-up but which also remain visibly contaminated. Incredibly, the pollution at one of these sites, Boobanabe, also in Kegbara Dere, dates back to a fire at a Shell oil well in 1970. One of village elders, Emadee Roberts Kpai, remembers the day when Shell first came to the area. "They promised that if they find oil here they'll transform our community and everybody will be happy." But the transformation was not what they had expected. Emadee's farm and fish-ponds have been devastated by oil pollution from nearby Shell pipelines. Now, he says "We have no hope for our children in this community". Whether Emadee's bleak assessment of the future comes true or not rests not just with Shell, but also with the Nigerian government. So far, its record of hold-ing oil companies to account has been woeful. President Muhammadu Buhari has pledged to restore the oil-wrecked environment, but his government has done little to implement the UN's recommendations from 2011. Doing so would be the best way of honouring the memory of Ken Saro-Wiwa.

Mark Dummett
London August 2018

Appendices

Ogoni flag signed by Ken Saro-Wiwa in 1994. Maynooth University Ken Saro-Wiwa Archive.

Chronology of key events in Ken Saro-Wiwa's life

10-Oct-41	Born in Bori, Ogoni
1958	Shell starts oil production in the Niger Delta
1961	Graduates from secondary school, Umuahia (SE Nigeria)
1962 – 65	BA (English) at University of Ibadan (SW Nigeria)
1964	First plays performed
1965 – 67	Teaches in Port Harcourt, Umuahia and Nsukka (SE Nigeria)
1967	Republic of Biafra declares independence; KSW moves to Lagos
1967	Administrator of Bonny Province (Rivers State) after Nigerian reconquest; Interim Advisory Council for Rivers State
1967 – 73	Teaching at University of Lagos
1968 – 73	Rivers State commissioner and member of Executive Council
1970	Nigerian Civil War (Biafran War) ends
1970	Ogoni chiefs petition against Shell-BP operations
1972 – 3	Commissioner for Ministry of Information and Home Affairs; removed for supporting Ogoni autonomy
1973	First two books (Tambari and Tambari in Dukana) published
1973	Sets up Saros International Limited
1985	Sets up Saros International Publishers
1985 – 90	Writes and produces Basi and Company
1986	Sozaboy published
1987 – 92	Accepts government appointments during supposed transition to democracy

1989 – 93	President of Ogoni Central Union
1989	On a Darkling Plain published
1990	MOSOP present Ogoni Bill of Rights
1990 – 93	President of Association of Nigerian Authors
1992 – 93	Repeated arrests
1993 – 95	President of Movement for the Survival of the Ogoni People
1993 – 95	Vice-chair of Unrepresented Nations and Peoples Organization
04-Jan-93	First Ogoni Day mobilises up to 60% of all Ogoni
1993	Military occupation of Ogoni
12-Jun-93	General elections; annulled by General Babangida
21-Jun-93	Arrested and detained until 22nd July; this is the period chronicled in A Month and A Day
1994	Awarded Right Livelihood Award and Fonlon-Nichols Prize
22-May-94	Arrested on trumped-up charges of involvement in death of 4 Ogoni leaders
1995	Awarded Goldman Environmental Prize
10-Nov-95	Executed with eight others following military-appointed tribunal
1996	Nominated posthumously for Nobel Peace Prize
1998	Death of General Abacha
1999	Return to civilian rule in Nigeria
2009	Shell settles out of court for $15.5 million on lawsuit brought by relatives of the "Ogoni Nine"

Select bibliography

This bibliography is intended as an introduction to the published work of Ken Saro-Wiwa. For more detailed bibliographic information, see Craig McLuckie and James Gibbs "An Annotated Bibliography" in Craig W. McLuckie and Aubrey McPhail. Eds. *Ken Saro-Wiwa: Writer and Political Activist* (Colorado: Lynne Rienner, 2000), pp. 245-284.

Some of Ken Saro-Wiwa's books

Non-Fiction

On a Darkling Plain: An Account of the Nigerian Civil War. Port Harcourt: Saros International, 1989.
Nigeria: The Brink of Disaster. Port Harcourt: Saros International, 1991.
Similia: Essays on Anomic Nigeria. London: Saros International, 1991.
Genocide in Nigeria: The Ogoni Tragedy. Port Harcourt: Saros International,

1992.

The Ogoni Nation Today and Tomorrow. Second Edition. Port Harcourt: Saros International, 1993.

"Closing Statement to the Military Appointed Tribunal." ALA *Bulletin*.21.4(1995): 3-4.

A Month and a Day: A Detention Diary. London: Penguin, 1995.

A Month and a Day and Letters. Foreword by Wole Soyinka. Oxford: Ayebia Clark Publishing Limited, 2005.

Note: *A Month and a Day* is a memoir recounting Saro-Wiwa's political activism and involvement with MOSOP up until his release from his first detention on 22nd July 1993. It was published during the year of his execution in 1995. A 2005 reprint entitled A *Month and a Day and Letters* includes the full text of A *Month and a Day* supported by other documents. These include some letters written by Saro-Wiwa during his second detention from 22nd May 1994 to 10th November 1995 as well as poems and tributes to Saro-Wiwa by writers and statespeople from around the world. One of the letters in that volume is reprinted in this volume. It is the author's last letter to Sr. Majella McCarron. Dated 14th September 1995, it is recognized as his last recorded letter. This volume places it into the context of the author's two year correspondence with McCarron which covers the period 20 October 1993 to 14 September 1995 that was not documented by the author in A *Month and a Day.*

Fiction

Sozaboy: A Novel in Rotten English Port Harcourt: Saros International Publishers, 1985. Reprinted London: Longman, 1994.

A Forest of Flowers Port Harcourt: Saros International, 1986. Reprinted Harlow: Longman, 1995.

Basi and Company: A Modern African Folktale Port Harcourt: Saros International, 1987.

Prisoner of Jebs Port Harcourt: Saros International, 1988.

Adaku and Other Stories Port Harcourt: Saros International, 1989.

The Singing Anthill: Ogoni Folktales Port Harcourt: Saros International, 1990.

Pita Dumbrok's Prison Port Harcourt: Saros International, 1991.

Lemona's Tale London: Penguin, 1996.

Television and Stage Plays

Basi and Company: Four Television Plays Port Harcourt: Saros International, 1987.

Four Farcical Plays London: Saros International, 1989.

Poetry

Songs in a Time of War Port Harcourt: Saros International, 1985.

Children's Books

Tambari Lagos: Longman, 1973.

Tambari in Dukana Lagos: Longman, 1979.

Mr. B Port Harcourt: Saros International, 1987.

Mr. B Saros Junior Series 2. London and Port Harcourt: Saros International, 1989.

Mr. B Goes to Lagos Port Harcourt: Saros International, 1989.

Mr. B Goes to the Moon Port Harcourt: Saros International, 1989.

A Bride for Mr. B Port Harcourt: Saros International, 1983.

Mr. B is Dead London: Saros International, 1991.

Mr. B's Mattress London: Saros International, 1992.

Segi Finds the Radio Port Harcourt: Saros International, 1991.

A Shipload of Rice London: Saros International, 1991.

The Transistor Radio (Port Harcourt: Saros International, 1989.

Key to the Maynooth University Ken Saro-Wiwa Archive

Letters from Ken Saro-Wiwa to Sr. Majella McCarron have been classified by Maynooth University University Library as follow:

PP/7/1	20-Oct-93
PP/7/2	01-Dec-93
PP/7/3	13-Jul-94
PP/7/4	EXACT DATE OF LETTER IS UNKNOWN. MCCARRON BELIEVES IT WAS WRITTEN IN JUNE OR JULY 1994
PP/7/5	24-Jul-94
PP/7/6	30-Jul-94
PP/7/7	15-Aug-94
PP/7/8	16-Sep-94
PP/7/9	01-Oct-94
PP/7/10	11-Oct-94
PP/7/11	19-Oct-94
PP/7/12	24-Oct-94
PP/7/13	27-Oct-94
PP/7/14	29-Oct-94

PP/7/15	29-Oct-94
PP/7/16	PHOTOCOPY OF THE FIRST TWO PAGES OF PP/7/15
PP/7/17	30-Oct-94
PP/7/18	22-Nov-94
PP/7/19	PHOTOCOPY OF PP/7/18
PP/7/20	24-Dec-94
PP/7/21	15-Jan-95
PP/7/22	07-Feb-95
PP/7/23	21-Mar-95
PP/7/24	24-Mar-95
PP/7/25	02-May-95
PP/7/26	19-Jun-95
PP/7/27	08-Jul-95
PP/7/28	14-Sep-95

The archives at Maynooth University

MAYNOOTH UNIVERSITY (formerly National University of Ireland Maynooth) is located 20 kilometres west of Dublin in Ireland's only university town. Irish universities are small by international standards, and the University has approximately 8,500 students and 26 academic Departments. There are two main library buildings—the John Paul II Library and the Russell Library (which holds the rare book and manuscript collections). The campus has recently undergone a major phase of expansion in research, teaching and service facilities. This expansion includes a major extension to the John Paul II Library, completed in 2013.

With a dedicated team of librarians, archivists and conservators, Maynooth University is ideally positioned to preserve, conserve and make available archives and special collections. The major extension to the Library means there is now custom-made space for the accommodation, consultation and exhibition of important archival collections, which conforms to best international standards. The conservation unit has attracted visitors nationally and internationally to view the conservation processes and to gain insights into best practice. Archival work on various collections has generated a great deal of interest in the holdings of the Library and the research opportunities that they offer. By cataloguing and preserving important archival collections, both print and electronic, researchers are now able to access material that had previously very limited availability. Collections are promoted via the library website; through a range of social media including Facebook, Flickr and the Russell Library blog and an ambitious programme of exhibitions. Many of the library exhibitions are open to the public and the University has a strong ethos of community engagement.

Both the John Paul II and Russell Libraries are shared with St. Patrick's College which was founded in 1795 (with the abolition of anti-Catholic penal legislation) as the headquarters of Ireland's Catholic hierarchy and its leading seminary. This makes Maynooth's libraries a key resource for the history of religion in Ireland in particular. Collections in the Russell Library span ten centuries and include medieval manuscripts, early printed books (the earliest printed in 1468) and works published by leading scholars across

Europe from the 16th to the mid-19th century, providing a superb treasury for scholarly research. The Maynooth University Library also has a central role in the Office of Public Works (OPW)-Maynooth University Archive and Research Centre at nearby Castletown House, Ireland's largest Palladian style house. This centre was established to facilitate the preservation and study of archives and other sources dealing with the history of Irish estates, their houses and inhabitants.

Recent significant archives donated to the Library include a selection of letters from Irish writer Seán Ó Faoláin to Brazilian academic Munira Hamud Mutran over a fourteen-year period and the literary papers of Teresa Deevy, one of the most prolific writers for the Irish Abbey Theatre in the 1930s. An exhibition of her work hosted by the Library in 2013 and the digitisation of the collection will go some way to re-establish Deevy alongside world-renowned Irish dramatists such as Sean O'Casey and J.M. Synge.

It is against this rich background that the letters of Ken Saro-Wiwa to Sister Majella McCarron join the holdings of Maynooth University. This is a particularly appropriate donation because of Maynooth University's long involvement with social movements in Ireland and abroad. St Patrick's College, for its part, has been central to the history of Irish missionary activity abroad. Like Sister Majella, many Irish missionaries worked in solidarity with local movements, or acted as development workers, particularly in health and education. It is to be hoped that this archive will be the first of many donations reflecting and recording this history.

To discuss donating archival material to the Library, please contact the University Librarian Cathal McCauley at Cathal.mccauley@mu.ie

More information about the Library can be found at https://www.maynoothuniversity.ie/library.

Resource toolkit

These are just some of the very wide range of online resources available for campaigners. Those listed here have very different themes and perspectives and are included here to help those new to the area to orient themselves.

News and information on development, environment, human rights, global justice

New Internationalist magazine http://newint.org/
Le Monde Diplomatique (in English) http://mondediplo.com/
ZNet progressive communications https://zcomm.org/znet/
Red Pepper magazine http://www.redpepper.org.uk/
Indigenous Environmental Network http://www.ienearth.org/
Rabble.ca news http://rabble.ca/
Daraja Press https://darajapress.com

Materials and discussion for campaigners

Trócaire campaigning resources http://www.trocaire.org/resources
into-ebooks NGO e-publisher http://www.into-ebooks.com/
Grassroots Global Justice Alliance http://ggjalliance.org/
Interface social movements journal http://www.interfacejournal.net/
Movement / Media Research Action Project http://mrap.info/
Comm-org online conference on community organising http://comm-org.wisc.edu/
MA in Community Education, Equality and Social Activism blog http://ceesa-ma.blogspot.ie/

Toolkits for effective movements for change

Activist Trauma Support https://www.activist-trauma.net/
Alliance of Community Trainers http://trainersalliance.org/
Learning for Sustainability http://learningforsustainability.net/
Organizing for Power http://organizingforpower.org/
Seeds for Change resources for activists http://www.seedsfor-change.org.uk/
Tools for Change training project http://www.toolsforchange.net/
Training for Change http://www.trainingforchange.org/
TRAPESE popular education collective http://trapese.clearerchannel.org/resources.php

Lightning Source UK Ltd.
Milton Keynes UK
UKHW02f2057011018
329820UK00011B/188/P